Molecular Biomarkers and More Efficient Therapies for Sepsis

Molecular Biomarkers and More Efficient Therapies for Sepsis

Guest Editors

Wen-Lin Su
Sheng-Kang Chiu
Chih-Hao Shen
Yi-Ting Chen

MDPI

Basel • Beijing • Wuhan • Barcelona • Belgrade • Novi Sad • Cluj • Manchester

Guest Editors

Wen-Lin Su
Division of Pulmonary and Critical Care Medicine
Taipei Tzu Chi Hospital
New Taipei City
Taiwan

Sheng-Kang Chiu
Department of Internal Medicine
Taipei Tzu Chi Hospital
New Taipei City
Taiwan

Chih-Hao Shen
Department of Internal Medicine
Tri-Service General Hospital
Taipei
Taiwan

Yi-Ting Chen
Department of Critical Care Medicine
Hualien Tzu Chi Hospital
Hualien City
Taiwan

Editorial Office
MDPI AG
Grosspeteranlage 5
4052 Basel, Switzerland

This is a reprint of the Special Issue, published open access by the journal *Biomedicines* (ISSN 2227-9059), freely accessible at: https://www.mdpi.com/journal/biomedicines/special_issues/JN440138PI.

For citation purposes, cite each article independently as indicated on the article page online and as indicated below:

Lastname, A.A.; Lastname, B.B. Article Title. *Journal Name* **Year**, *Volume Number*, Page Range.

ISBN 978-3-7258-3869-1 (Hbk)
ISBN 978-3-7258-3870-7 (PDF)
https://doi.org/10.3390/books978-3-7258-3870-7

Contents

biomedicines

Editorial

Molecular Biomarkers and More Efficient Therapies for Sepsis

Wen-Lin Su [1,2,*], Sheng-Kang Chiu [2,3], Chih-Hao Shen [4] and Yi-Ting Chen [2,5]

1 Division of Pulmonary and Critical Care Medicine, Taipei Tzu Chi Hospital, Buddhist Tzu Chi Medical Foundation, No. 289, Jianguo Rd., Xindian Dist., New Taipei City 23142, Taiwan
2 School of Medicine, Tzu Chi University, No. 701, Sec. 3, Chung Yang Rd., Hualien City 970, Taiwan; csk33kimo@hotmail.com (S.-K.C.); kateytc@gmail.com (Y.-T.C.)
3 Division of Infectious Diseases, Department of Internal Medicine, Taipei Tzu Chi Hospital, Buddhist Tzu Chi Medical Foundation. No. 289, Jianguo Rd., Xindian Dist., New Taipei City 23142, Taiwan
4 Division of Pulmonary and Critical Care Medicine, Department of Internal Medicine, Tri-Service General Hospital, National Defense Medical Center No. 325, Sec. 2, Cheng-Kung Road, Neihu 114, Taipei 11490, Taiwan; potato652@yahoo.com.tw
5 Department of Critical Care Medicine, Hualien Tzu Chi Hospital, Buddhist Tzu Chi Medical Foundation, No. 707, Sec. 3, Chung Yang Rd., Hualien City 970, Taiwan
* Correspondence: williamsu2007@gmail.com

1. Introduction

Sepsis remains a leading cause of morbidity and mortality worldwide, representing a substantial burden on healthcare systems [1–3]. Furthermore, sepsis is an ill-defined syndrome with an immunopathophysiology characterized by simultaneous hyperinflammation and immune suppression, which makes it challenging to address varying immune statuses at different stages of sepsis [4]. Nanotechnology represents a promising trend in the future diagnosis and management of sepsis [5]. Recent advancements in molecular biomarkers and innovative therapeutic strategies offer promising avenues for addressing the multifaceted challenges of sepsis. This editorial synthesizes the key findings from recent studies (Figure 1), providing a roadmap for future research and clinical applications.

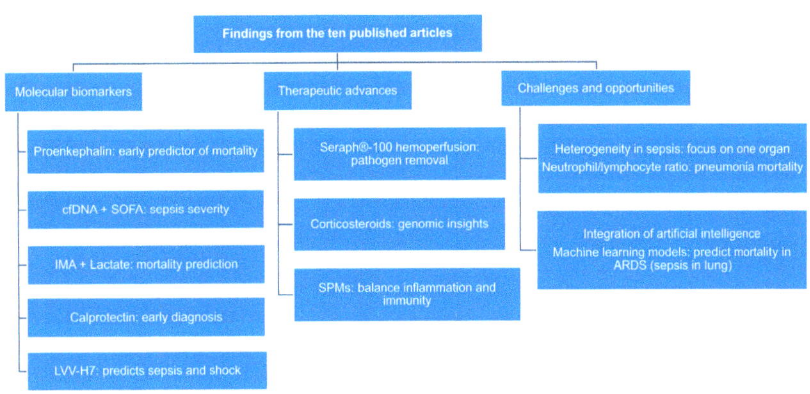

Figure 1. This updated figure exclusively reflects the findings derived from the ten published articles. The insights are categorized as molecular biomarkers and therapeutic advances.

Received: 26 January 2025
Accepted: 8 February 2025
Published: 14 February 2025

Citation: Su, W.-L.; Chiu, S.-K.; Shen, C.-H.; Chen, Y.-T. Molecular Biomarkers and More Efficient Therapies for Sepsis. *Biomedicines* **2025**, *13*, 468. https://doi.org/10.3390/biomedicines13020468

2. Advances in Molecular Biomarkers

2.1. Proenkephalin as an Early Predictor of Mortality

Emerging biomarkers such as proenkephalin (PENK) have shown potential in predicting in-hospital mortality among septic shock patients, demonstrating strong correlations with metabolic and inflammatory biomarkers [6]. PENK's utility underscores the importance of integrating point-of-care diagnostics in early sepsis management.

2.2. Cell-Free DNA (cfDNA) Combined with SOFA in Severity Assessment

cfDNA, including nuclear and mitochondrial components, has been identified as a marker for sepsis severity [7]. Its combination with clinical scores such as SOFA enhances diagnostic precision and prognostication. This underscores the potential of multiplex biomarker panels in stratifying sepsis risk.

2.3. Ischemia-Modified Albumin (IMA) and Lactate in Mortality Prediction

The combination of IMA and lactate levels provides a robust tool for predicting mortality in septic shock patients [8]. This synergistic approach exemplifies how biomarkers can complement each other for enhanced predictive accuracy.

2.4. Calprotectin in Early Diagnosis of Infections

Calprotectin's ability to predict bacterial infections offers a cost-effective solution for early sepsis detection, reducing ICU stays and mortality [9]. This health economic analysis highlights calprotectin's economic and clinical benefits as part of early intervention strategies.

2.5. LVV-Hemorphin-7 (LVV-H7) in Predicting Sepsis

LVV-H7, a metabolite of cell-free hemoglobin catalyzed by cathepsins D and G during infection, shows potential for predicting sepsis and shock in critically ill patients with acute changes in SOFA scores [10]. These results highlight the utility of cell-free hemoglobin metabolites in sepsis prognostication.

3. Innovative Therapeutic Approaches

3.1. Seraph®-100 Hemoperfusion

Novel therapies such as the Seraph®-100 hemoperfusion device have demonstrated efficacy in removing bacterial pathogens in a simulated hemoperfusion study, representing a breakthrough in extracorporeal therapies for sepsis-related bacteremia [11].

3.2. Corticosteroids and Genomic Insights

Corticosteroids, long debated for their role in sepsis management, may benefit from genomic and transcriptomic stratification to identify responsive subgroups [12]. These therapies modulate the immune response, stabilize the cardiovascular system, and potentially facilitate organ restoration. This personalized approach could optimize outcomes while minimizing risks.

3.3. Specialized Pro-Resolving Mediators (SPMs)

In sepsis, severe inflammation occurs early, followed by paradoxical immunosuppression in later stages. SPMs offer a dual advantage by resolving inflammation without inducing immunosuppression [13]. Their role in managing the immunosuppressive phase of sepsis provides a balanced therapeutic strategy.

4. Challenges and Opportunities

4.1. Heterogeneity in Sepsis

Sepsis is not a singular disease but a syndrome with diverse etiologies and manifestations. Biomarker-driven phenotyping can aid in tailoring therapies to individual patient profiles, addressing this heterogeneity. Pneumonia, a major cause of sepsis, may benefit from predictors such as the neutrophil/lymphocyte ratio and pneumonia severity index for mortality risk assessment [14].

4.2. Integration of Artificial Intelligence

Machine learning and imaging algorithms are emerging as powerful tools for sepsis diagnosis and monitoring. These technologies can complement molecular biomarkers to refine risk stratification and therapeutic decisions. For example, integrating P/F ratios and chest X-ray data in machine learning models may help predict mortality in SARS-CoV-2-associated ARDS [15].

5. Future Directions

As illustrated in Figure 2, the future directions focus on personalized medicine and combinatorial therapies for heterogeneous sepsis, including variations in early versus late sepsis and different infection sources. Economic evaluation and global collaboration should also be prioritized, driven by artificial intelligence to tackle this complex disease.

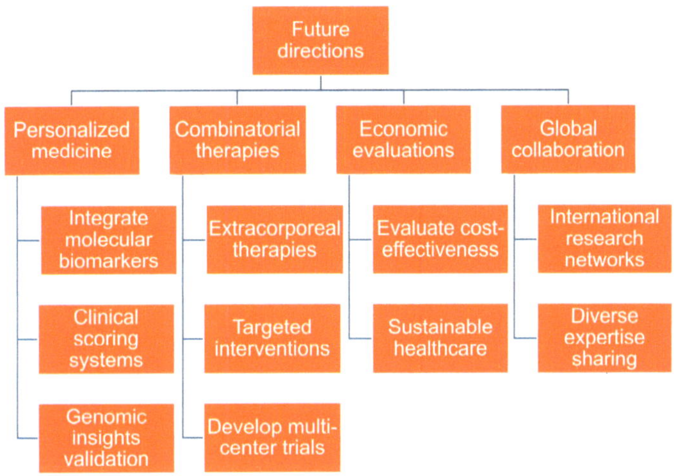

Figure 2. This figure illustrates the future directions in sepsis biomarkers and therapies. It highlights key themes—Personalized Medicine, Combinatorial Therapies, Economic Evaluations, and Global Collaboration—along with their associated strategies.

5.1. Personalized Medicine

The integration of clinical scoring systems [16] combined with molecular biomarkers [17] such as proteomics [18] and genomic insights [19] paves the way for personalized sepsis management. The different pathogens and their drug resistance and virulence may have different outcomes, such as mortality [20]. In addition, the varying definitions of infection in critically ill populations [21] and the definition of sepsis may also offer important predictions of mortality outcomes [22]. Future research should focus on validating biomarker panels in diverse populations.

5.2. Combinatorial Therapies

The sepsis treatment guidelines, as established by the Surviving Sepsis Campaign in 2021 [23], still lack strong evidence for molecular therapies. Combining extracorporeal therapies [24–27], targeted pharmacological interventions [28,29], and supportive care [30,31] can maximize treatment efficacy. As many multi-center trials have begun of targeted pharmacological therapy in sepsis [32,33], additional multi-center trials are needed to establish best practices for such combinatorial approaches.

5.3. Economic Evaluations

Given the financial burden of sepsis, cost-effectiveness analyses should be integral to the development and deployment of new biomarkers and therapies [34]. This ensures the sustainability of healthcare interventions.

5.4. Global Collaboration

Sepsis is a global health challenge. International collaborations can accelerate the discovery and implementation of effective solutions, leveraging diverse expertise and resources [35].

6. Conclusions

The convergence of molecular biomarkers, innovative therapies, and advanced technologies heralds a new era in sepsis management. By addressing the complexities of this syndrome through personalized and evidence-based approaches, we can improve outcomes for millions of patients worldwide.

Author Contributions: Conceptualization, W.-L.S.; writing—original draft preparation, W.-L.S.; writing—review and editing, S.-K.C., C.-H.S., Y.-T.C. and W.-L.S.; project administration, S.-K.C. and W.-L.S.; funding acquisition, W.-L.S. All authors have read and agreed to the published version of the manuscript.

Funding: This work was supported by grants from the Buddhist Tzu Chi Medical Foundation [TCMF-CP 112-02] and Taipei Tzu Chi Hospital [TCRD-TPE-114-09]. The funding body had no role in the conceptualization or design of the study; data collection; analysis; or the preparation or decision to publish the manuscript.

Conflicts of Interest: The authors declare no conflicts of interest.

References

1. van den Berg, M.; van Beuningen, F.E.; ter Maaten, J.C.; Bouma, H.R. Hospital-related costs of sepsis around the world: A systematic review exploring the economic burden of sepsis. *J. Crit. Care* **2022**, *71*, 154096. [CrossRef] [PubMed]
2. Paoli, C.J.; Reynolds, M.A.; Sinha, M.; Gitlin, M.; Crouser, E. Epidemiology and Costs of Sepsis in the United States-An Analysis Based on Timing of Diagnosis and Severity Level. *Crit. Care Med.* **2018**, *46*, 1889–1897. [CrossRef]
3. Rudd, K.E.; Kissoon, N.; Limmathurotsakul, D.; Bory, S.; Mutahunga, B.; Seymour, C.W.; Angus, D.C.; West, T.E. The global burden of sepsis: Barriers and potential solutions. *Crit. Care* **2018**, *22*, 232. [CrossRef] [PubMed]
4. Wiersinga, W.J.; van der Poll, T. Immunopathophysiology of human sepsis. *EBioMedicine* **2022**, *86*, 104363. [CrossRef] [PubMed]
5. Pant, A.; Mackraj, I.; Govender, T. Advances in sepsis diagnosis and management: A paradigm shift towards nanotechnology. *J. Biomed. Sci.* **2021**, *28*, 6. [CrossRef] [PubMed]
6. Verras, C.; Bezati, S.; Bistola, V.; Ventoulis, I.; Matsiras, D.; Tsiodras, S.; Parissis, J.; Polyzogopoulou, E. Point-of-Care Serum Proenkephalin as an Early Predictor of Mortality in Patients Presenting to the Emergency Department with Septic Shock. *Biomedicines* **2024**, *12*, 1004. [CrossRef]
7. de Miranda, F.S.; Claudio, L.; de Almeida, D.S.M.; Nunes, J.B.; Barauna, V.G.; Luiz, W.B.; Vassallo, P.F.; Campos, L.C.G. Cell-Free Nuclear and Mitochondrial DNA as Potential Biomarkers for Assessing Sepsis Severity. *Biomedicines* **2024**, *12*, 933. [CrossRef]
8. Jin, B.Y.; Lee, S.; Kim, W.; Park, J.H.; Cho, H.; Moon, S.; Ahn, S. Ischemia-Modified Albumin, Lactate, and Combination for Predicting Mortality in Patients with Septic Shock in the Emergency Department. *Biomedicines* **2024**, *12*, 1421. [CrossRef]
9. Havelka, A.; Larsson, A.O.; Mårtensson, J.; Bell, M.; Hultström, M.; Lipcsey, M.; Eriksson, M. Analysis of Calprotectin as an Early Marker of Infections Is Economically Advantageous in Intensive Care-Treated Patients. *Biomedicines* **2023**, *11*, 2156. [CrossRef] [PubMed]
10. Wu, Y.K.; Chung, H.W.; Chen, Y.T.; Chen, H.C.; Chen, I.H.; Su, W.L. Association of LVV-Hemorphin-7 with Sepsis and Shock: Roles of Cathepsin D and G in Hemoglobin Metabolism in a Prospective ICU Cohort Study. *Biomedicines* **2024**, *12*, 2789. [CrossRef]

11. Lacquaniti, A.; Smeriglio, A.; Campo, S.; La Camera, E.; Lanteri, G.; Giunta, E.; Monardo, P.; Trombetta, D. In Vitro Simulated Hemoperfusion on Seraph(®)-100 as a Promising Strategy to Counteract Sepsis. *Biomedicines* **2024**, *12*, 575. [CrossRef]
12. Lazar, A. Recent Data about the Use of Corticosteroids in Sepsis-Review of Recent Literature. *Biomedicines* **2024**, *12*, 984. [CrossRef]
13. Padovani, C.M.; Yin, K. Immunosuppression in Sepsis: Biomarkers and Specialized Pro-Resolving Mediators. *Biomedicines* **2024**, *12*, 175. [CrossRef]
14. Tekin, A.; Wireko, F.W.; Gajic, O.; Odeyemi, Y.E. The Neutrophil/Lymphocyte Ratio and Outcomes in Hospitalized Patients with Community-Acquired Pneumonia: A Retrospective Cohort Study. *Biomedicines* **2024**, *12*, 260. [CrossRef]
15. Cysneiros, A.; Galvão, T.; Domingues, N.; Jorge, P.; Bento, L.; Martin-Loeches, I. ARDS Mortality Prediction Model Using Evolving Clinical Data and Chest Radiograph Analysis. *Biomedicines* **2024**, *12*, 439. [CrossRef] [PubMed]
16. Bhargava, A.; López-Espina, C.; Schmalz, L.; Khan, S.; Watson, G.L.; Urdiales, D.; Updike, L.; Kurtzman, N.; Dagan, A.; Doodlesack, A.; et al. FDA-Authorized AI/ML Tool for Sepsis Prediction: Development and Validation. *NEJM AI* **2024**, *1*, AIoa2400867. [CrossRef]
17. Mi, Y.; Burnham, K.L.; Charles, P.D.; Heilig, R.; Vendrell, I.; Whalley, J.; Torrance, H.D.; Antcliffe, D.B.; May, S.M.; Neville, M.J.; et al. High-throughput mass spectrometry maps the sepsis plasma proteome and differences in patient response. *Sci. Transl. Med.* **2024**, *16*, eadh0185. [CrossRef]
18. Miao, H.; Chen, S.; Ding, R. Evaluation of the Molecular Mechanisms of Sepsis Using Proteomics. *Front. Immunol.* **2021**, *12*, 733537. [CrossRef] [PubMed]
19. Hernandez-Beeftink, T.; Guillen-Guio, B.; Lorenzo-Salazar, J.M.; Corrales, A.; Suarez-Pajes, E.; Feng, R.; Rubio-Rodríguez, L.A.; Paynton, M.L.; Cruz, R.; García-Laorden, M.I.; et al. A genome-wide association study of survival in patients with sepsis. *Crit. Care* **2022**, *26*, 341. [CrossRef] [PubMed]
20. Umemura, Y.; Ogura, H.; Takuma, K.; Fujishima, S.; Abe, T.; Kushimoto, S.; Hifumi, T.; Hagiwara, A.; Shiraishi, A.; Otomo, Y.; et al. Current spectrum of causative pathogens in sepsis: A prospective nationwide cohort study in Japan. *Int. J. Infect. Dis.* **2021**, *103*, 343–351. [CrossRef] [PubMed]
21. Calandra, T.; Cohen, J.; FRCP for the International Sepsis Forum Definition of Infection in the ICU Consensus Conference. The International Sepsis Forum Consensus Conference on Definitions of Infection in the Intensive Care Unit. *Crit. Care Med.* **2005**, *33*, 1538–1548. [CrossRef] [PubMed]
22. Rhee, C.; Jones, T.M.; Hamad, Y.; Pande, A.; Varon, J.; O'Brien, C.; Anderson, D.J.; Warren, D.K.; Dantes, R.B.; Epstein, L.; et al. Prevalence, Underlying Causes, and Preventability of Sepsis-Associated Mortality in US Acute Care Hospitals. *JAMA Netw. Open* **2019**, *2*, e187571. [CrossRef]
23. Evans, L.; Rhodes, A.; Alhazzani, W.; Antonelli, M.; Coopersmith, C.M.; French, C.; Machado, F.R.; McIntyre, L.; Ostermann, M.; Prescott, H.C.; et al. Surviving sepsis campaign: International guidelines for management of sepsis and septic shock 2021. *Intensive Care Med.* **2021**, *47*, 1181–1247. [CrossRef] [PubMed]
24. Bottari, G.; Ranieri, V.M.; Ince, C.; Pesenti, A.; Aucella, F.; Scandroglio, A.M.; Ronco, C.; Vincent, J.L. Use of extracorporeal blood purification therapies in sepsis: The current paradigm, available evidence, and future perspectives. *Crit. Care* **2024**, *28*, 432. [CrossRef] [PubMed]
25. Ronco, C.; Chawla, L.; Husain-Syed, F.; Kellum, J.A. Rationale for sequential extracorporeal therapy (SET) in sepsis. *Crit. Care* **2023**, *27*, 50. [CrossRef] [PubMed]
26. Zhang, L.; Feng, Y.; Fu, P. Blood purification for sepsis: An overview. *Precis. Clin. Med.* **2021**, *4*, 45–55. [CrossRef] [PubMed]
27. Kang, J.H.; Super, M.; Yung, C.W.; Cooper, R.M.; Domansky, K.; Graveline, A.R.; Mammoto, T.; Berthet, J.B.; Tobin, H.; Cartwright, M.J.; et al. An extracorporeal blood-cleansing device for sepsis therapy. *Nat. Med.* **2014**, *20*, 1211–1216. [CrossRef]
28. Tindal, E.W.; Armstead, B.E.; Monaghan, S.F.; Heffernan, D.S.; Ayala, A. Emerging therapeutic targets for sepsis. *Expert. Opin. Ther. Targets* **2021**, *25*, 175–189. [CrossRef] [PubMed]
29. Zhang, Y.-y.; Ning, B.-t. Signaling pathways and intervention therapies in sepsis. *Signal Transduct. Target. Ther.* **2021**, *6*, 407. [CrossRef] [PubMed]
30. Mer, M.; Schultz, M.J.; Adhikari, N.K. Core elements of general supportive care for patients with sepsis and septic shock in resource-limited settings. *Intensive Care Med.* **2017**, *43*, 1690–1694. [CrossRef] [PubMed]
31. Perner, A.; Rhodes, A.; Venkatesh, B.; Angus, D.C.; Martin-loeches, I.; Preiser, J.-C.; Vincent, J.-L.; Marshall, J.; Reinhart, K.; Joannidis, M.; et al. Sepsis: Frontiers in supportive care, organisation and research. *Intensive Care Med.* **2017**, *43*, 496–508. [CrossRef]
32. François, B.; Lambden, S.; Fivez, T.; Gibot, S.; Derive, M.; Grouin, J.M.; Salcedo-Magguilli, M.; Lemarié, J.; De Schryver, N.; Jalkanen, V.; et al. Prospective evaluation of the efficacy, safety, and optimal biomarker enrichment strategy for nangibotide, a TREM-1 inhibitor, in patients with septic shock (ASTONISH): A double-blind, randomised, controlled, phase 2b trial. *Lancet Respir. Med.* **2023**, *11*, 894–904. [CrossRef] [PubMed]

33. Laterre, P.F.; Pickkers, P.; Marx, G.; Wittebole, X.; Meziani, F.; Dugernier, T.; Huberlant, V.; Schuerholz, T.; François, B.; Lascarrou, J.B.; et al. Safety and tolerability of non-neutralizing adrenomedullin antibody adrecizumab (HAM8101) in septic shock patients: The AdrenOSS-2 phase 2a biomarker-guided trial. *Intensive Care Med.* **2021**, *47*, 1284–1294. [CrossRef] [PubMed]
34. Afshar, M.; Arain, E.; Ye, C.; Gilbert, E.; Xie, M.; Lee, J.; Churpek, M.M.; Durazo-Arvizu, R.; Markossian, T.; Joyce, C. Patient Outcomes and Cost-Effectiveness of a Sepsis Care Quality Improvement Program in a Health System. *Crit. Care Med.* **2019**, *47*, 1371–1379. [CrossRef]
35. Llitjos, J.F.; Carrol, E.D.; Osuchowski, M.F.; Bonneville, M.; Scicluna, B.P.; Payen, D.; Randolph, A.G.; Witte, S.; Rodriguez-Manzano, J.; François, B. Enhancing sepsis biomarker development: Key considerations from public and private perspectives. *Crit. Care* **2024**, *28*, 238. [CrossRef]

biomedicines

Review

Immunosuppression in Sepsis: Biomarkers and Specialized Pro-Resolving Mediators

Cristina M. Padovani * and Kingsley Yin

Department of Cell Biology and Neuroscience, Rowan-Virtua School of Translational Biomedical Engineering and Sciences, Virtua Health College of Life Sciences of Rowan University, Stratford, NJ 08084, USA; yinki@rowan.edu
* Correspondence: padova29@rowan.edu

Abstract: Severe infection can lead to sepsis. In sepsis, the host mounts an inappropriately large inflammatory response in an attempt to clear the invading pathogen. This sustained high level of inflammation may cause tissue injury and organ failure. Later in sepsis, a paradoxical immunosuppression occurs, where the host is unable to clear the preexisting infection and is susceptible to secondary infections. A major issue with sepsis treatment is that it is difficult for physicians to ascertain which stage of sepsis the patient is in. Sepsis treatment will depend on the patient's immune status across the spectrum of the disease, and these immune statuses are nearly polar opposites in the early and late stages of sepsis. Furthermore, there is no approved treatment that can resolve inflammation without contributing to immunosuppression within the host. Here, we review the major mechanisms of sepsis-induced immunosuppression and the biomarkers of the immunosuppressive phase of sepsis. We focused on reviewing three main mechanisms of immunosuppression in sepsis. These are lymphocyte apoptosis, monocyte/macrophage exhaustion, and increased migration of myeloid-derived suppressor cells (MDSCs). The biomarkers of septic immunosuppression that we discuss include increased MDSC production/migration and IL-10 levels, decreased lymphocyte counts and HLA-DR expression, and increased GPR18 expression. We also review the literature on the use of specialized pro-resolving mediators (SPMs) in different models of infection and/or sepsis, as these compounds have been reported to resolve inflammation without being immunosuppressive. To obtain the necessary information, we searched the PubMed database using the keywords sepsis, lymphocyte apoptosis, macrophage exhaustion, MDSCs, biomarkers, and SPMs.

Keywords: macrophages; lymphocytes; apoptosis; infection; MDSCs; biomarkers; sepsis; SPMs

Citation: Padovani, C.M.; Yin, K. Immunosuppression in Sepsis: Biomarkers and Specialized Pro-Resolving Mediators. *Biomedicines* **2024**, *12*, 175. https://doi.org/10.3390/biomedicines12010175

Academic Editor: Wen-Lin Su

Received: 21 December 2023
Revised: 10 January 2024
Accepted: 11 January 2024
Published: 13 January 2024

1. Introduction

During infection, the immune system initiates a complex and tightly coordinated response not only to clear the host of the infection, but also to ensure it will be prepared the next time it encounters the same pathogen. The host's immune system utilizes a highly regulated balance of pro-inflammatory, anti-inflammatory, and resolution processes in order to achieve this. If one of these processes becomes dysregulated, the delicate homeostatic balance will become disrupted, ultimately leading to chronic disease and/or death.

Sepsis is defined as a life-threatening infection that can lead to organ failure. It is the leading cause of death in U.S. hospitals, and its high mortality rate may be attributed to a lack of diagnostics and viable treatment regimens. Sepsis management typically focuses on blunting the early, hyper-proinflammatory phase of sepsis, where the patient's tissues and organs are overwhelmed by a release of cytokines—a phenomenon known as cytokine release syndrome (CRS). CRS is a clinical syndrome that comprises the early phase of sepsis. CRS can be defined as a hyper-proinflammatory immune response to a stimulus leading to elevated levels of pro-inflammatory cytokines, inflammation, and organ dysfunction [1]. There is significant overlap between this definition and the consensus definition of the early phase of sepsis; however, this early phase typically is followed by a late phase of immunosuppression [1]. While it is reported that states of severe immunosuppression predispose a patient to developing CRS, this association does not appear to always work in the reverse [1]. There is suggestion that CRS and sepsis could be linked, but this connection has not been completely confirmed [1,2].

Historically, there was very little focus on management of the late, immunosuppressive phase of sepsis. In this stage, the patient's immune system is considered to be hypoactive, as these deaths are attributed to the patient being unable to clear the original infection or to the acquisition of a secondary infection [3]. Better treatments and strategies for sepsis management are necessary; however, before effective modalities targeting the late phase of sepsis can be implemented, there needs to be a better understanding of the factors contributing to sepsis-induced immunosuppression.

Here, we review current knowledge of immunosuppression in sepsis. We describe three major hallmarks of sepsis-induced immunosuppression (lymphocyte apoptosis, monocyte/macrophage exhaustion, and myeloid-derived suppressor cell migration), with a particular focus on the dysregulation of immune cell responses. In addition to this, we compare two highly debated phenomena that both appear to occur at least in part in septic patients: endotoxin tolerance and monocyte/macrophage exhaustion.

At present, there is no one treatment that has consistently been shown to reduce morbidity and mortality in sepsis patients, despite decades of research and hundreds of clinical trials [4]. Even though there have been recent advancements in antibiotic therapy, there is still a need for new, effective treatments for sepsis. We argue that successful treatment in sepsis patients relies on physician understanding of the patient's immune status, which is highly dependent on what stage of sepsis the patient is in, and what therapeutics would therefore be the most effective. These therapeutics rely on biomarkers that may enable real-time immunological profiling of individual sepsis patients, especially those in the late, immunosuppressive phase. We will review accepted and potential biomarkers that could aid physicians in delineating their patients' immune status during late sepsis.

2. Methods

2.1. Search Term Strategy

From October 2023 to January 2024, research studies were identified by searching the research database PubMed. The following string of search terms was used to identify peer-reviewed articles in this database: "immunosuppression" AND "sepsis"; "sepsis" AND "monocytes OR macrophages"; "sepsis" AND "lymphocyte apoptosis"; "sepsis" AND "myeloid-derived suppressor cells OR MDSCs"; "sepsis" AND "biomarkers"; sepsis AND "specialized pro-resolving mediators OR SPMs OR Resolvin OR Lipoxin". A reference search was conducted using the final selected articles.

2.2. Inclusion Criteria

Types of studies. The studies that are included contained mostly mixed data research, primarily composed of hypothesis-driven benchtop research, prospective cohort studies, and observational studies. Only articles that were written in the English language were evaluated.

Types of participants. Reviewed studies included human participants who met the following criteria: males and females; patients who were diagnosed with sepsis according to that hospital's standards (typically by using sepsis-3 criteria); living in the U.S., not including U.S. territories; patients without an active diagnosis of sepsis (control group). The racial background and age of participants were not used as selection criteria. Reviewed studies also included preclinical animal models of sepsis and infection.

2.3. Exclusion Criteria

While we tried to include mostly articles that were published within the past 10 years, no research articles were excluded on the basis of published year alone.

2.4. Data Extraction

Data extracted from the final selected articles included article information, methodology, and results of the studies. Information was collected on the article such as title, authors, and published year. Data on methodology included demographics of the study population (if applicable), study design, sample size, qualitative aspects such as tissue appearance for mouse experiments, and quantitative aspects such as patient laboratory results and statistical significance.

2.5. Data Analysis

Extracted data were analyzed quantitatively and descriptively. A review of the results was run by the researchers to provide statistical significance across references.

3. The Innate Immune System and Inflammatory Response in Infection

Levels of defense against pathogens include host barriers and the human immune system [5]. Host barriers include physical, chemical, and biological obstacles that help prevent the entrance of pathogens. Physical barriers include the skin and mucous membranes, chemical barriers include pH and lysozymes, and biological barriers include the host's normal flora [5]. If these barriers are breached, then the pathogen will encounter the host's immune system. The human immune system can be subdivided into two systems: the innate immune system and the adaptive immune system. The innate immune response is quick-acting, does not discriminate, and does not develop memory, whereas the adaptive immune response is slow-acting, is highly specific, and develops memory (See Table 1 and Figure 1) [6].

Table 1. Innate vs. Adaptive Immune Systems.

	Innate Immune System	Adaptive Immune System
Response Speed?	Quick	Slow upon first exposure
Nonspecific or Specific?	Nonspecific	Specific
Development of Traditional Immunologic Memory?	No	Yes

The innate immune response acts first and is usually strong enough to clear an infection without triggering the adaptive immune response. A key feature of this immune system is that it must be able to differentiate between "self" and "non-self" so that it does not accidentally attack the body's own cells, thinking they are pathogens [6]. It is able to perform such discrimination via recognition of pathogen-associated molecular patterns (PAMPs), conserved motifs found on a pathogen's surfaces, which help the immune system identify the pathogen as foreign, or "non-self" [6]. Some examples of PAMPs are lipopolysaccharide (LPS) found in the outer membranes of gram-negative bacteria, peptidoglycans found in gram-negative and gram-positive bacterial cell walls, lipoteichoic acid from gram-positive bacteria, single-stranded DNA and double-stranded RNA found in some viruses, and beta-glucans found in fungal cell walls [6]. The innate immune system contains a series of pattern recognition receptors (PRRs), proteins that are able to recognize PAMPs; once activated, these PRRs elicit a series of cell signaling events that activate the cell's specific effector functions that will be helpful to rid the host of the invading pathogen—namely, via the inflammatory response [6].

The inflammatory response is a crucial part of the body's innate immune system, and when it acts acutely, it is very effective at clearing an infection. The cardinal signs of inflammation include redness, heat, swelling, and pain [9], and these are clinical manifestations of the protection offered by the innate immune system. The acute inflammatory response is dominated by neutrophils, white blood cells whose primary goals are to phagocytose the invading pathogen and to release pro-inflammatory mediators [10]. They are able to migrate to the site where the pathogen breached the first level of defense following the release of local mediators, such as histamine and nitric oxide, which function to increase blood vessel vasodilation [11]. Macrophages are another type of white blood cell that are also recruited to the site of injury to assist with pathogen removal. Pro-inflammatory mediators are produced via activation of several signaling pathways, the most prevalent of which is the nuclear factor kappa light chain enhancer of activated B cells (NF-κB) pathway [12]. The pro-inflammatory mediators produced include cytokines, chemokines, free radicals, and inflammatory lipid mediators (e.g., prostaglandins, leukotrienes) (See Figure 2). This pro-inflammatory response functions to clear pathogens; however, if this response persists for too long, then the tissues themselves can become damaged.

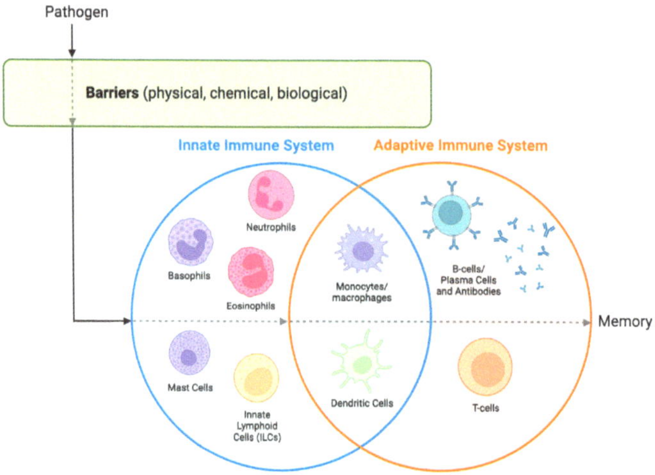

Figure 1. Barriers of defense. Once a pathogen breaches the first line of defense (host barriers), it is then encountered by the innate immune system. Main cell players in the innate immune system include neutrophils, basophils, eosinophils, mast cells, and innate lymphoid cells (ILCs; e.g., natural killer cells) [5–7]. If the innate immune system is not successful at completely clearing this pathogen, the pathogen needs to contend with the adaptive immune system. Main cell players in the adaptive immune system include B cells, plasma cells, and T cells [5]. Monocytes/macrophages and dendritic cells are innate immune cells that are said to bridge both the innate and adaptive immune systems, as they are the main initiators of the adaptive immune system [8]. Once the adaptive immune system successfully removes the pathogen, immunological memory is established. Key: defense breach is depicted by a dotted arrow. Created in BioRender.com.

Maintenance of the innate immune system relies on immunological homeostasis, whereby the host's pro- and anti-inflammatory responses balance one another in order to help return the host to baseline [13]. Therefore, resolution circuits are activated in order to ensure that this response and other aspects of the innate immune response act acutely and do not persist, and these circuits function to reduce neutrophil activation and increase macrophage efferocytosis of apoptotic neutrophils [14,15]. This is a critical step in the acute inflammatory response—if this inflammatory response either persists for too long or fails to clear the pathogen, chronic inflammation can and will ensue [10]. Because the innate immune system is always active within the host, it is typically not inhibited, but rather, it is regulated to help return the host to homeostasis; however, there are some ways in which the host can ensure its cessation as a backup mechanism [16]. For example, in addition to increasing IL-10 levels and the activity of anti-inflammatory T-cell subsets such as T-regulatory cells (Tregs), which both function to dampen the pro-inflammatory response, the innate immune system also seemingly modulates PRR expression on innate immune cells that favor inhibition rather than activation [16]. These inhibitory PRRs (iPRRs) are said to fine-tune the degree of innate immune cell activation via inhibition of cell signaling pathways [16].

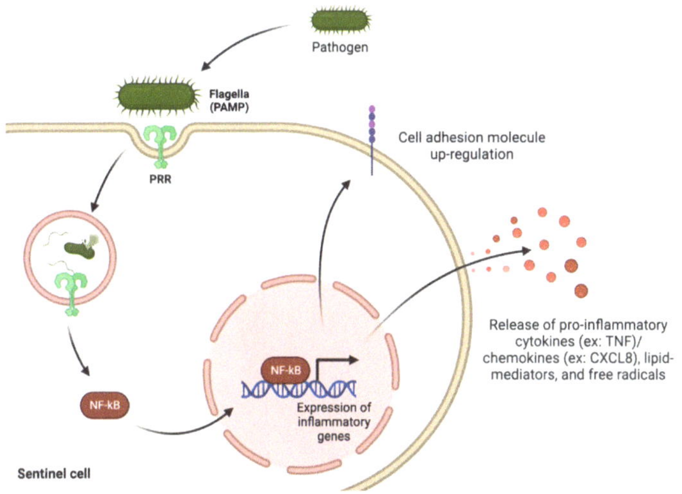

Figure 2. The innate immune system. Activation of the innate immune system begins with sentinel cell recognition of non-self. This usually occurs in the form of pathogen-associated molecular patterns (PAMPs)–pattern recognition receptor (PRR) interactions. Many downstream signaling cascades can become triggered by this PAMP–PRR interaction (which are cell- and pathogen-specific), but one of the most important signaling pathways that becomes activated is the nuclear factor kappa light chain enhancer of activated B cells (NF-κB) pathway. The effector molecules produced via this pathway are pro-inflammatory cytokines/chemokines, lipid mediators, and free radicals. Created in BioRender.com.

If this innate inflammatory process is not successful at eliminating the body of the infection, then the adaptive immune response will become activated. The adaptive immune response primarily comprises B and T lymphocytes and memory functions. The activation of these lymphocytes leads to the production of antibodies (B cells) and increased macrophage signaling to clear pathogens (T cells).

4. The Adaptive Immune System and Response to Infection

The adaptive immune system is the last line of defense against invading pathogens. This system is triggered directly by T cell recognition of foreign invaders or by presentation of foreign antigen to T cells by one of the three antigen-presenting cells (APCs; monocytes/macrophages, dendritic cells, and B cells) [11]. APCs contain major histocompatibility complex (MHC) class II (MHC-II) on their surfaces, and these molecules are required for MHC-restricted antigen presentation to T cells and the subsequent activation of the adaptive immune response [17,18]. From here, the activated T cells can help stimulate other immune cells (including other types of T cells), can activate B cells to produce class-switched antigen-specific antibodies, and can clear the invading pathogen [11]. Finally, completion of this response allows for the development of immunological memory.

Overactivation of the adaptive immune system is commonly implicated in the development of autoimmune disease [19]. This is likely because the adaptive immune response is robust and produces memory. With that being said, there are many regulatory molecules in place to ensure proper functioning and activation (i.e., that T and B cells will not be inappropriately activated). One such mechanism occurs during T- and B-cell development: positive and negative selection (i.e., central tolerance) [20]. These selection processes occur during development to ensure that T and B cells both recognize foreign antigens but do not recognize foreign antigens too strongly [20]. Once these lymphocytes "pass" these selection processes, they are able to circulate in the body; however, there are more selection mechanisms in place in the periphery that the host subjects the naïve lymphocytes to (i.e., peripheral tolerance) [20]. For example, in order to ensure that activation of a naïve T lymphocyte is warranted, this activation process in the periphery requires two signals (See Figure 3).

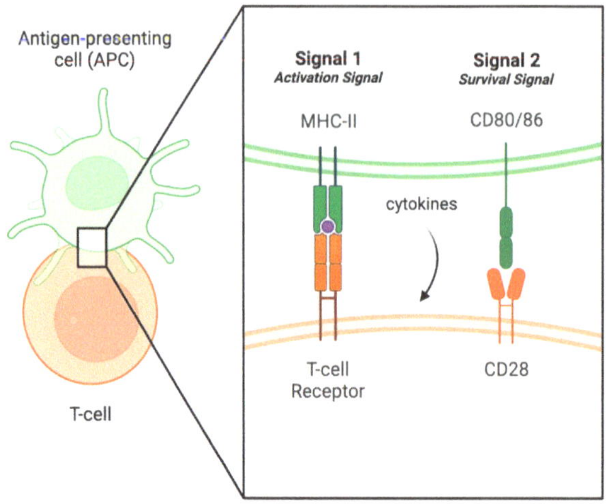

Figure 3. Peripheral tolerance and T cell activation. Activation of the adaptive immune system can begin with T-cell activation by an antigen-presenting cell (APC). This figure highlights the important concept of peripheral tolerance, or the two-signal hypothesis. Created in BioRender.com.

The first signal is the "activation signal", which comprises the interaction of the T-cell receptor (TCR) on the T cell with an MHC-II molecule on an APC [21]. The second signal is the "survival signal", which comprises the interaction of the CD28 molecule on the T cell with a co-stimulatory molecule (CD80/86) on the APC [21]. Once the T cell becomes activated, it is able to carry out its effector functions, depending on what is needed at that time [21]. Typically, the T cell is "told" what is needed at that time for pathogen clearance via cytokines released by the APC onto the T cell [21]. These effector functions include the additional release of cytokines, activation of B cells, direct destruction of the pathogen, or activation of other cells or T-cell subtypes [21]. Additionally, some white blood cells have molecules expressed on their surfaces known as "immune regulatory molecules" or "checkpoints", which can modulate T-cell functions [22]. These regulatory molecules are cytotoxic T-lymphocyte-associated protein 4 (CTLA-4), programmed cell death protein 1 (PD-1), and programmed death ligand 1 (PD-L1) [22]. CTLA-4 and PD-1 are both located on the surfaces of T cells and act as halting mechanisms of T-cell activation and function [23]. CTLA-4 interacts strongly with APC co-stimulatory molecules (i.e., signal two molecules), and this interaction inhibits T-cell activation and functioning [23]. Similarly, PD-1 can interact with PD-L1 on the APC's surface, leading to T-cell anergy and exhaustion [23].

The adaptive immune response is maintained throughout the course of the host's lifetime via the development of immunological memory. Following an adaptive immune response, memory T cells and memory B cells are created; upon repeated exposure to the same antigen in the future, these memory cells will be able to quickly and efficiently clear the pathogen [8].

5. Inflammation Resolution

Many studies have disproved the idea that resolution of inflammation (i.e., inflammation resolution) is a passive process, whereby the mere cessation of the inflammatory response is adequate enough to return the body to homeostasis [24]. It is now widely accepted that the resolution of inflammation is an active process, whereby the body synthesizes endogenous pro-resolving compounds and activates or suppresses certain cells/mediators, ultimately resolving inflammation without being immunosuppressive [25]. These pro-resolving compounds are known as specialized pro-resolving mediators (SPMs) [25], and they will be described in more detail toward the end of this article.

The goal of inflammation resolution is to allow the body's tissues that have been harmed by a stimulus (e.g., pathogen, damaged tissue) or by the acute inflammatory response to the stimulus to return to homeostasis. Inflammation resolution is an active, coordinated process whereby certain molecular/cellular mechanisms take place in various immune cells in order to heal the harmed tissues [24]. One key event in this process is the recruitment and activation of macrophages to the site of infection [24]. As mentioned above, another key event in this process is the release of endogenous SPMs. Expectedly, the loss of or the dysregulation of inflammation resolution mechanisms results in pathological chronic inflammation and disease [24].

6. Mechanisms of Immunosuppression in Sepsis

Three hallmarks of sepsis-induced immunosuppression include: monocyte/macrophage exhaustion, lymphocyte apoptosis, and myeloid-derived suppressor cell migration (See Figure 4).

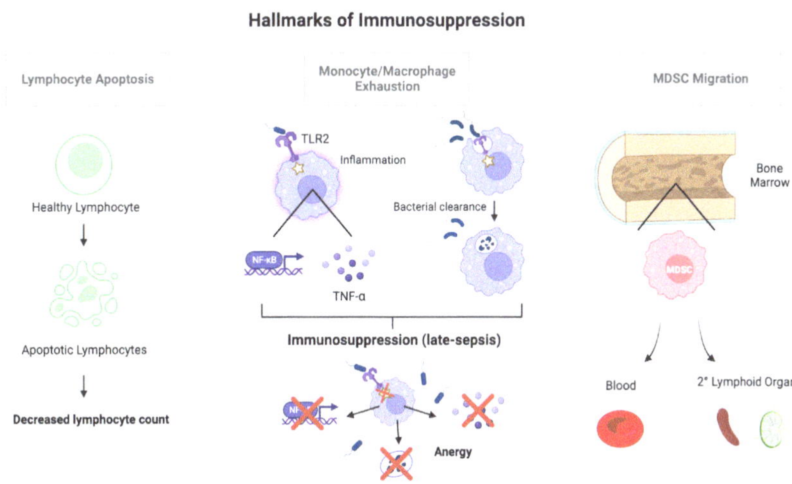

Figure 4. Mechanisms of sepsis-induced immunosuppression. Illustration of three major mechanisms underlying the pathophysiology of sepsis-induced immunosuppression: lymphocyte apoptosis, monocyte/macrophage exhaustion, and migration of myeloid-derived suppressor cells. Key: TLR2: Toll-like receptor 2; MDSC: myeloid-derived suppressor cell. Created in BioRender.com.

6.1. Monocyte/Macrophage Exhaustion

One hallmark of sepsis-induced immunosuppression is monocyte/macrophage exhaustion [3]. Monocytes are circulating white blood cells derived from the bone marrow and are major players in the innate immune system. Their primary role is to detect changes in homeostasis (usually caused by an infection and/or tissue damage) and respond to these changes by replenishing the pool of macrophages that are present in the tissues [26]. Once monocytes become activated, they turn into effector cells and produce cytokines, present antigens to adaptive immune cells, phagocytose, efferocytose, initiate inflammation, or resolve inflammation [26]. When monocytes initiate tissue inflammation, they exit the blood and enter the tissues at the site of the inflammation, thereby becoming macrophages. Macrophages can either be recruited to a specific tissue (which is what happens during the inflammatory response), or they can exist as residents within the body's tissues. Macrophages can polarize into different subtypes depending on which subtype can best assist the host at that time. The previously accepted polarization classes are M0, M1-like, and M2-like; an M0 macrophage is a nonactivated tissue resident macrophage, an M1-like macrophage is a pro-inflammatory macrophage, and an M2-like macrophage is an anti-inflammatory macrophage (See Figure 5) [26]. It is classically understood that M1-like macrophages have increased production of pro-inflammatory cytokines, enhanced antigen presentation, and enhanced phagocytosis capabilities [27]. M2-like macrophages have diminished pro-inflammatory cytokine output and weakened antigen presentation, but enhanced phagocytosis capabilities [27,28]. These phenotypes can be artificially induced in vitro via monocyte stimulation, with LPS and IFN-γ leading to the development of a "classic" M1-like macrophage and IL-4 leading to the development of a "classic" M2-like macrophage [29,30].

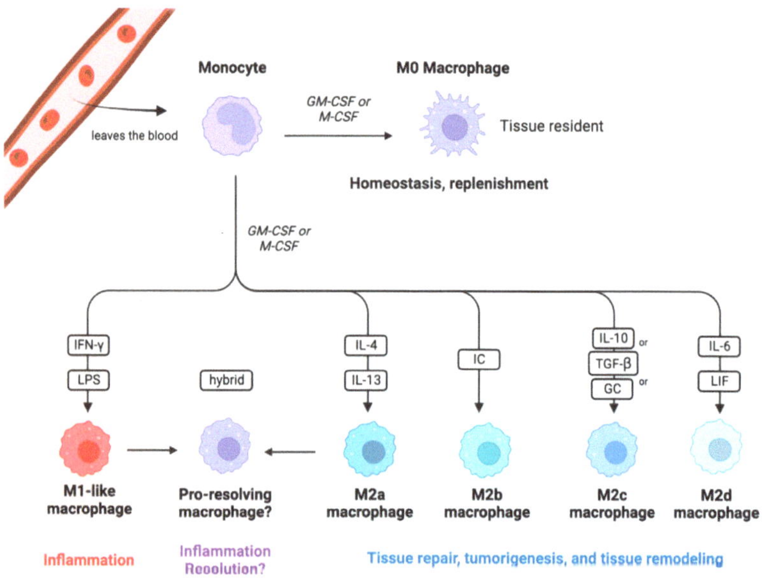

Figure 5. Monocyte polarization. This figure highlights the classical monocyte polarization scheme and classes of macrophages. Each class has a unique role within the immune system. Key: GM-CSF: granulocyte macrophage colony-stimulating factor; M-CSF: macrophage colony-stimulating factor; LPS: lipopolysaccharide; IC: immune complex; GC: glucocorticoid; LIF: leukemia inhibitory factor. Created in BioRender.com.

M1-like macrophages are stimulated by IFN-γ and LPS; they are key initiators of the inflammatory response and are critical for the removal of bacteria [31]. On the other hand, M2-like macrophages are highly phagocytic and are key drivers of tissue remodeling following damage, tumor progression via the production of angiogenic factors, and parasite removal [31]. M2-like macrophages are also a main driver in the development of allergies. Additionally, the M2-like macrophage subtype can be further subclassified into M2a, M2b, M2c, and M2d [32,33]. These subtypes vary in what cytokines they release, what surface markers they express, and what specific roles they play within M2-like macrophage responses. M2a macrophages are stimulated via IL-4 and IL-13; they enhance cell growth and tissue repair and enhance macrophage endocytosis [32,33]. M2b macrophages are stimulated via immune complexes (IC); they are able to regulate the host's immune status by secreting both pro- and anti-inflammatory cytokines [32,33]. M2b macrophages are also well known for their ability to promote tumor progression [33]. M2c macrophages are stimulated by IL-10, TGF-β, or glucocorticoids (GCs); like M2b macrophages, they are highly immune-regulatory but specialize in efferocytosis and phagocytosis of other apoptotic cells [32,33]. M2d macrophages are stimulated via IL-6 and leukemia inhibitory factor (LIF); they promote angiogenesis and tumor progression via the release of vascular endothelial growth factors and dampening of the immune response [32,33].

While the M1/M2 paradigm is helpful for investigators studying these different capabilities of macrophages in vitro, many researchers find that limiting the description of macrophage phenotypes to only M1-like or M2-like is not clinically relevant, as this tidy polarization process does not occur at a tissue level [29,30,34–36]. Current research explains that activated macrophages have high levels of plasticity and do not solely polarize into an M1-like or M2-like phenotype at the tissue level; rather, they adopt a phenotype that is highly dependent on both the environment they are in and their own ontogeny [29,30]. One such newly reported phenotype is the "pro-resolving" macrophage phenotype [29,30,37,38]. These dynamic, pro-resolving macrophages function primarily to resolve inflammation so that the host can return to homeostasis, while simultaneously contributing to wound/tissue repair [29,30]. Interestingly, these pro-resolving macrophages are said to have qualities of the classic M1- and M2-like phenotypes, with aspects that contribute to both pro-inflammatory and anti-inflammatory pathways [37]. Some studies even report that abnormal activation of this pro-resolving phenotype can lead to chronic organ dysfunction, highlighting their direct importance in the inflammation resolution process [39–41].

6.1.1. PD-1 and PDL-1

Monocytes/macrophages contribute to sepsis-induced immunosuppression in a number of ways. Monocytes/macrophages can modulate the expression of their immune regulatory markers, an action that has implications for other immune cells and physiological pathways [42]. Changes in the expression of immune regulatory markers such as CTLA-4, PD-1, and PD-L1 can alter these interactions [22]. PD-1 (also referred to as CD279) is a transmembrane protein located on the surfaces of many immune cells; it is expressed in the highest concentration on activated T cells [43]. When PD-1 interacts with its ligand PD-L1, the T cell becomes anergic, or unresponsive to viable antigens [44–47]. PD-L1 is expressed by certain APCs (such as monocytes/macrophages) in inflammatory scenarios as well as by tumor cells [44]. Additionally, certain pro-inflammatory cytokines (especially IFN-γ) can trigger various signal transduction cascades within APCs, leading directly to increased expression of PD-L1 on their surfaces [48].

Physiologically, the PD-1/PD-L1 pathway aids in controlling the degree of inflammation at the tissue level by deactivating T cells that are directly contributing to the inflammation so as to protect tissues from immune-mediated damage [44]. Tumor cells overexpress PD-L1 in order to exist for longer periods of time without being "seen" by the adaptive immune system [44,48]. The PD-1/PD-L1 pathway has therefore become a major contributor to immunosuppression within the tumor microenvironment in many different types of cancers (notably melanoma and small-cell lung cancer) [49]. Understanding this major contribution of PD-1/PD-L1 to cancer-induced immunosuppression has led to the development of a new class of cancer immunotherapies called checkpoint inhibitors [22,49]. Due to many parallels existing between cancer-induced immunosuppression and sepsis-induced immunosuppression [3], checkpoint inhibitors were tested in clinical trials of sepsis patients and cecal ligation and puncture (CLP) sepsis models [50,51]. Hotchkiss et al. was the first to study the effects and safety profile of the anti-PD-1 immunotherapy agent nivolumab in a small clinical trial of sepsis patients [50]. More recently, Yang et al. showed that blocking PD-L1 on monocytes led to a dramatic increase in survival of CLP-septic mice [51].

Additionally, many researchers have assessed the potential correlation between immune regulatory marker expression and outcomes due to sepsis. Ruan et al. found that there is an increased number of cells expressing both PD-1 and PD-L1 in CLP-septic mice [52]. This finding was also true for humans, as there are reported increases in expression of both PD-1 and PD-L1 in human septic patients [52–55]. While changes in immune regulatory marker expression on monocytes/macrophages contributes to sepsis-induced immunosuppression, this finding is not specific to septic patients. As previously mentioned, cancer cells can implement similar changes in regulatory marker expression levels in order to evade the immune system. A more explicit example of monocytes/macrophages contributing to immunosuppression due to sepsis is monocyte/macrophage exhaustion [3].

6.1.2. Exhausted Monocytes/Macrophages

Exhausted monocytes/macrophages are "hypoactive", meaning that they have a decreased capacity to perform their normal effector functions when stimulated [56–59]. These normal effector functions include presenting antigens via HLA-DR, phagocytosing, efferocytosing, and releasing pro-inflammatory cytokines (e.g., TNF-α, IL-6, IL-8) via the NF-κB pathway [26]. Therefore, NF-κB status is an accepted marker of exhaustion [60]. As monocytes/macrophages from septic patients also have similarly reduced capacities of these normal functions, it can be inferred that septic monocytes/macrophages are exhausted [61]. A recent study revealed that septic macrophages had decreased pro-inflammatory cytokine production in response to various ligands, including bacteria and LPS [62]. In this study, the authors found that whole blood samples taken from 61 septic patients (blood taken within 24 h of meeting sepsis-3 criteria) had an overwhelming presence of inflammation, yet leukocytes from these samples had depressed responses to both LPS and bacteria [62], providing evidence that septic leukocytes were similar to the "exhausted" phenotype. This hypoactive phenotype present in septic monocytes/macrophages can be recapitulated in vitro by repetitively challenging monocytes/macrophages with isolated, low-dose bacterial endotoxins such as LPS [58,63].

It is important to briefly discuss here that monocyte/macrophage exhaustion is different from the phenomenon called endotoxin tolerance (ET). ET has many similar characteristics to exhaustion [61], and it is defined as reduced immune cell functional responsiveness to the bacterial endotoxin LPS after previously encountering LPS [56,57]. Blackwell et al. examined ET via monocyte cytokine production; they discovered that tolerized monocytes had impaired activation of NF-κB [64], which is also observed in exhaustion. Tolerized monocytes/macrophages exhibited a reduced capacity to react to subsequent LPS challenge; primarily of note was the drastic decrease in NF-κB-mediated TNF-α production upon LPS restimulation [65–70]. While the monocytes from late-septic patients seem to have many aspects similar to monocytes/macrophages experiencing ET, the conventional understanding of ET does not encompass the fundamental characteristics of immunosuppression observed in septic patients [59]. For example, monocytes/macrophages experiencing ET have enhanced bacterial phagocytosis, as it has been theorized that ET monocytes/macrophages adopt a phenotype similar to M2-like macrophages [56]. However, exhausted monocytes/macrophages and monocytes from septic patients demonstrate impaired bacterial phagocytosis, even though the mechanism underlying this decline is not known [71,72]. For these reasons, the emerging literature more accurately describes septic monocytes/macrophages as exhausted rather than tolerized [59]. However, the mechanism(s) underlying monocyte/macrophage exhaustion during the immunosuppressive stage of sepsis remains elusive.

Several possibilities for sepsis-induced monocyte/macrophage exhaustion include increased production of the anti-inflammatory cytokine IL-10, down-regulation of the NF-κB signaling pathway, decreased HLA-DR expression, and reduced lymphocyte signaling due to lymphocyte cell death [57,61,70]. Another possibility is the high mobility group box 1 (HMGB1) protein, which has been shown to increase macrophage pyroptosis [73–75], decrease macrophage efferocytosis ability [76], and enhance the suppressive activity of an immunosuppressive cell population called myeloid-derived suppressor cells (MDSCs) [77].

To overcome monocyte/macrophage anergy (i.e., exhaustion) several research teams have used IFN-γ in order to stimulate monocyte activity [78]. In one study performed by

Döcke et al., the authors were able to increase HLA-DR expression in anergic monocytes, leading to sepsis resolution in eight of nine septic patients [78]. On the other hand, a more recent study showed that increased plasma IFN-γ levels were associated with secondary *Candida* sp. infections in late-sepsis patients [79]. Surprisingly, the authors also showed that IFN-γ reduced macrophage phagocytosis of zymosan particles [79]. Together, these two contradictory studies suggest that IFN-γ cannot be reliably regarded as an appropriate treatment to reverse sepsis-induced monocyte/macrophage exhaustion.

6.2. Lymphocyte Apoptosis

Another hallmark of sepsis-induced immunosuppression is lymphocyte apoptosis [3]. Before apoptosis can be discussed in more detail, it is important to mention that there are numerous types of regulated and unregulated cell death processes involving complex signaling cascades. The regulated and clinically relevant processes include, but are not limited to, autophagy, ferroptosis, pyroptosis, necroptosis, and apoptosis (See Figure 6). Autophagy is a process of regulated cell death by which autophagosomes collect cell waste products and deliver them to lysosomes for destruction [80]. This process is mainly employed to maintain homeostasis within the cell and to manage lipid metabolism [81]. Ferroptosis is a process of regulated cell death and is primarily caused by the accumulation of iron [82]. This accumulation leads to the development of reactive oxygen species (ROS), which contributes to both cell membrane rupture and lipid peroxidation [81]. Pyroptosis is a process of regulated cell death whereby destruction of the cell membrane is triggered by the activation of a multi-protein complex called the inflammasome [82,83]. Necroptosis is a process of regulated cell death that eventually leads to an influx of ions, swelling of the cell, and eventual breakdown of the membrane [82]. Due to many similarities between necroptosis and apoptosis (apoptosis is described in detail below), it has been suggested that necroptosis exists as a backup mechanism for apoptosis [81].

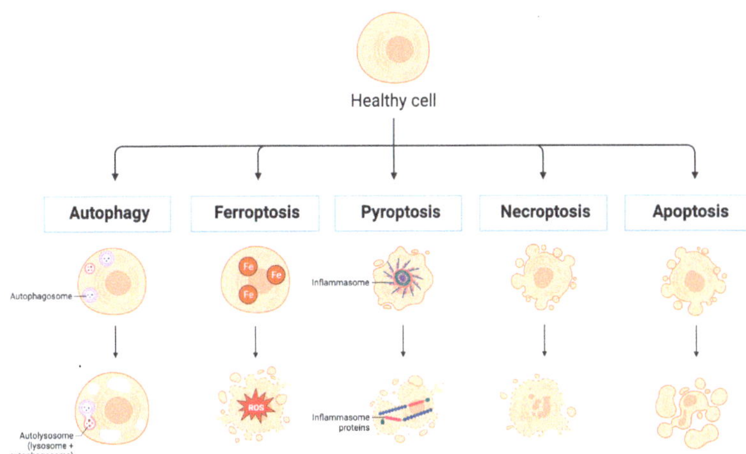

Figure 6. Mechanisms of regulated cell death. This figure highlights some important processes of regulated cell death: autophagy, ferroptosis, pyroptosis, necroptosis, and apoptosis. Key: Fe: iron; ROS: reactive oxygen species. Created in BioRender.com.

Apoptosis, or programmed cell death, is an active cellular process that is essential for proper functioning and regulation of many cells within the body, including immune cells. Apoptosis of faulty lymphocytes during and following lymphocyte development is crucial for proper immune functioning. There are two pathways that lead to lymphocyte apoptosis: "death by neglect" and "death by instruction" (See Figure 7) [84]. As discussed earlier, there are many mechanisms in place during lymphocyte development and activation to ensure proper performance and homeostasis within the population. If a lymphocyte fails to meet certain functional requirements, it will die via apoptosis ("death by neglect"). On the other hand, if a lymphocyte functions too aggressively, it will die via apoptosis ("death by instruction"). These apoptosis pathways that regulate lymphocyte development and expansion are necessary to ensure there are no self-reactive and/or hyporesponsive lymphocytes present in the immune system's repertoire.

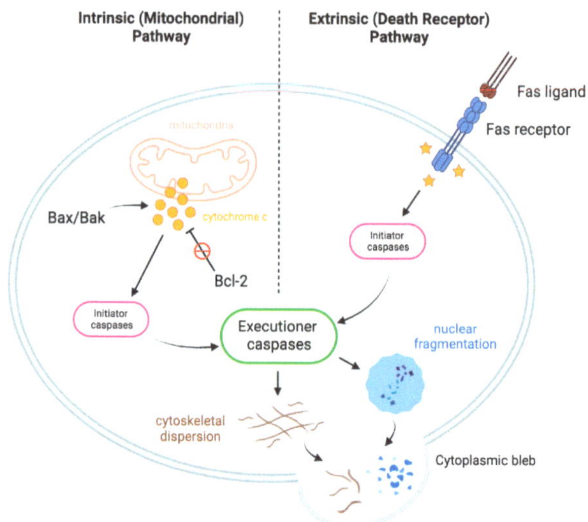

Figure 7. Apoptosis pathways. This figure demonstrates two pathways of apoptosis: the intrinsic (mitochondrial) pathway and the extrinsic (death receptor) pathway. Created in BioRender.com.

A defining feature of apoptosis is that the programmed cell death event does not trigger inflammation: the process occurs neatly and "quietly." This highly coordinated pathway involves two main groups of players that work with each other in concert to ensure that this organized process remains balanced: apoptosis activators and apoptosis regulators. There are two main pathways of apoptosis activation: the extrinsic pathway and the intrinsic pathway [85]. Both pathways involve an apoptotic signal, which activates a proteinase called an initiator caspase, which in turn activates another proteinase called an executioner caspase [85]. The two pathways differ in how the initiator caspase becomes activated, but the end result remains the same. The extrinsic pathway, also known as the *death receptor pathway*, involves the Fas ligand (FasL) binding to a cell's Fas receptor; the intrinsic pathway, also known as the *mitochondrial pathway*, involves the mitochondrial release of cytochrome c into the cytoplasm, which carries out effector functions necessary for apoptosis completion [86]. Two proteins that assist in the activation of apoptosis (and hence are considered "pro-apoptotic" proteins) are Bax and Bak, and their interaction with each other triggers destabilization of the mitochondrial membrane (via oligomerization) and the subsequent release of cytochrome c into the cytoplasm [85,87]. If Bax and Bak do not interact/oligomerize, the mitochondrial membrane remains stabilized, and there is no release of cytochrome c; if Bax and Bak do oligomerize, the membrane becomes destabilized, leading to the release of cytochrome c into the cytoplasm and completion of apoptosis [87]. Equally as important as apoptosis activation is the regulation of apoptosis by proteins within the Bcl-2 family. Bcl-2 is a regulatory protein within this family that functions as a direct regulator of Bax and Bak. Bcl-2 sequesters Bax and Bak so that they cannot oligomerize; this function earned Bcl-2 its "anti-apoptotic" title [86]. The level of Bcl-2, therefore, is a good biomarker to study changes in apoptosis within cells.

While some or all of the processes may be occurring in some way during the sepsis spectrum, they are difficult to definitively test for in a clinical setting. It is clear that cell death mechanisms are occurring in sepsis patients, as one of the main features of late sepsis is a decreased white blood cell count, and more specifically, lymphopenia [88]. For the purposes of this review, we will specifically focus on lymphocyte apoptosis, as that has been studied the most in experimental settings of late sepsis (i.e., in CLP sepsis models).

While lymphocyte apoptosis is a normal physiologic process necessary for proper immune functioning (i.e., a crucial part of the central tolerance of lymphocytes), this process can become dysregulated or can be hijacked by microbes, leading to an unnecessary loss of lymphocytes. Microbes can induce apoptosis in lymphocytes and other types of leukocytes in order to artificially create a more permissive environment for their survival within the host cells [89]. By triggering apoptosis within lymphocytes, microbes are able to thrive in the body because there are fewer immune cells capable of clearing them from the host. For example, *Bacillus anthracis*, the causative agent of anthrax, releases lethal toxin (LTx), which can lead to direct caspase release in macrophages, deceptively triggering their cell death pathways [90]. In the CLP model of polymicrobial sepsis, there is ~10% increase in the number of apoptotic lymphocytes compared with control mice [89]. Studies have shown that increased expression of Bcl-2 leads to increased survival of CLP-septic mice by ~40–50%, specifically through apoptosis prevention in splenic lymphocytes [89,91,92]. Additionally, there are numerous accounts confirming apoptosis of lymphocytes in CLP-septic mice using specific transferase-mediated dUTP nick-end labeling (TUNEL) staining methods [93–95].

Another mechanism of lymphocyte suppression that falls outside of, but is closely related to, the apoptotic pathways is the PD-1/PD-L1 pathway that was described earlier. While this pathway does not lead to apoptosis of lymphocytes, it renders the lymphocytes inactive, producing a very similar outcome. While the application of checkpoint inhibitors (i.e., PD-1 inhibitors) for sepsis-induced immunosuppression is still in its nascent stages, the enhancement of lymphocyte functionality in late sepsis may provide an exciting avenue for researchers moving forward.

In an attempt to target sepsis-induced immunosuppression, there have been several clinical trials whereby lymphocytes were targeted for restoration. In one study, recombinant GM-CSF or G-CSF was administered to late-septic patients in order to increase their lymphocyte counts [96]. These attempts initially seemed promising, but the beneficial effects did not persist long-term. In a similar clinical trial, IL-7 was administered to late-septic patients in an attempt to increase their populations of T lymphocytes, but the patients did not improve long-term [97]. In both of these instances, the failures were attributed to pathological migration and activation of myeloid-derived suppressor cells (MDSCs) [98]. It was postulated that MDSCs suppressed T lymphocyte activity and proliferation, negating the beneficial action of the intervention.

6.3. Myeloid-Derived Suppressor Cells

The third hallmark of sepsis-induced immunosuppression is increased myeloid-derived suppressor cell (MDSC) production/migration [3]. MDSCs are a group of heterogenous, immature cells of the myeloid lineage that work to suppress the innate and adaptive immune systems [99]. MDSCs can be divided into two main populations based on their lineage (although there are others that have been reported [100]): monocytic MDSCs (M-MDSCs) and polymorphic MDSCs (PMN-MDSCs) [99,101,102]. Both M-MDSCs and PMN-MDSCs are able to exert their inhibitory effects on the innate and adaptive immune systems in similar ways, with the primary mechanisms being via the release of ROS, production of arginase-1, and release of the anti-inflammatory cytokine IL-10 [99]. MDSCs exist in the bone marrow and are triggered to migrate into the blood and secondary lymphoid tissues via chemokine CXCL2 and IL-8 gradients [99]. Overall, MDSCs function as immunosuppressive cells, but their exact role and the meaning behind their migration patterns in patients with sepsis remains to be seen.

There is some controversy regarding the significance of the MDSC migration that is observed in septic patients [103–106]. Some studies report that the increase in MDSC migration out of the bone marrow in sepsis patients is partly responsible for the adverse clinical outcomes due to late sepsis [98,106–108]. In fact, Tang et al. suggests that inhibiting MDSCs would be one of the most effective treatments for sepsis-induced immunosuppression overall [108]. In septic patients, having elevated numbers of MDSCs in the peripheral blood was associated in some patients with longer ICU hospital stays and a higher statistical likelihood of acquiring a secondary infection [98]. In LPS-induced immunosuppression, MDSCs migrate from the bone marrow to the blood/secondary lymphoid tissues and inhibit the proliferation of lymphocytes [107]. Additional studies have shown that not only are MDSCs directly inhibitory on lymphocytes, but also their production of arginase-1 further disrupts T-cell functions [109]. Similarly, MDSCs can indirectly cause immunosuppression by activating Tregs [98]. This anti-inflammatory T-cell subset functions to inhibit other innate and adaptive immune cells, primarily via the release of IL-10 [98].

Conversely, there are also reports that MDSC migration during late sepsis could actually contribute to long-term survival in sepsis patients [103,104,106,110,111]. This paradoxical function of MDSCs was studied extensively in CLP-sepsis models. In one study, MDSCs were adoptively transferred into CLP mice early on in the infection timeline, and these mice had decreased mortality compared with sham animals [103]. In another study, mouse MDSC expansion was inhibited by pretreatment with the chemotherapeutic gemcitabine, and survival to CLP was dramatically reduced in the gemcitabine treatment group compared with the sham animals [104]. Similarly, in another study, inhibition of CLP mouse MDSC populations via GR-1 neutralization led to decreased survival compared with the sham animals [111]. In another infection model system, in vivo depletion of MDSCs in mice infected with *Trypanosoma cruzi* (the causative agent of Chagas disease) led to increased mortality compared with the sham animals [110].

A study by Schrijver et al. provides a strong case for the expansion and migration of M-MDSCs in particular correlating with improved survival in sepsis patients [112]. In this study, blood was taken from patients in eight ICUs with pneumonia secondary to sepsis [112]. The authors found that patients with high levels of M-MDSCs overall had significantly reduced 90-day mortality rates and improved survival compared with patients with high levels of PMN-MDSCs [112]. The contention is that due to the heterogeneity of MDSC populations, perhaps overall, MDSCs play a dual role during sepsis, with each unique MDSC population evolving over time after sepsis onset [103,112,113].

Overall, in sepsis patients, there is an elevation in MDSC numbers in the blood and the secondary lymphoid tissues. While a simple explanation would be that the migrating MDSCs are acting as the trigger for sepsis-induced immunosuppression, perhaps what is actually occurring here is that the host is attempting to correct the early hyper-proinflammatory phase of sepsis with MDSC migration. Due to unidentified dysregulation pathways in sepsis patients, perhaps the recruitment of MDSCs is not robust enough in these particular patients; therefore, the hyper-proinflammatory response persists until the late stage of sepsis sets in, and their presence then accentuates the immunosuppression. MDSC expansion and migration in sepsis appears to be a highly dynamic process, and these cells, conceivably, are malleable to the infection at hand. Although the complete consequences of increased MDSC production in sepsis have not been fully elucidated, their presence in the blood as a biomarker in the immunosuppressive phase of sepsis cannot be disputed.

7. Biomarkers

One of the most difficult aspects of developing sepsis therapeutics is the heterogeneity of the disease. There have been decades of failed sepsis therapeutics in clinical trials that were based on promising preclinical research. The discrepancy lies in the understanding that there are many factors contributing to patient outcomes due to sepsis, and the challenge in moving toward successful interventions stems from the variability of the immune system's actions. The way the immune system responds to infection can be influenced by the stage of the inflammatory timeline in which it is acting. Therefore, the therapeutics administered in sepsis rely heavily on specific biomarkers, which allow physicians to decipher the immune status of their patients. Then, the physicians can tailor treatments more accurately and, in a way, more individually. For example, while common practice is to administer antibiotics in order to prevent detrimental outcomes due to sepsis, studies have shown that the antibiotics must be administered within the first hour of the sepsis diagnosis to provide any perceived benefit [114,115]; otherwise, administration of broad-spectrum antibiotics could actually facilitate the development of multi-drug-resistant infections, potentially causing an infection epidemic within the hospital [114]. Part of understanding the lack of favorable outcomes in sepsis due to antibiotic administration is appreciating that while microbial infections trigger sepsis, the main driver of sepsis is the patient's own immune system—not the infection itself. Because sepsis itself is a clinical diagnosis (i.e., there is no "one test" that can diagnose sepsis), many times it either goes undiagnosed or it is diagnosed outside of the window of perceived benefit of antibiotic administration: once this window passes, the risk of antibiotic administration (i.e., antibiotic resistance) outweighs potential benefits [114]. The need for new, effective sepsis therapeutics will therefore depend heavily on quick thinking by the physician to test for specific biomarkers that could shed light on the patient's immune status, as well as what therapeutic would best be warranted for administration at that time. While there likely will not be a single treatment that will be equally effective for every single patient with sepsis, an ideal treatment would be one that supports the immune system's intrinsic ability to combat the pathogen while also resolving the infectious inflammation before it further causes damage to the body [116].

While there is no one test that definitively tells the physician that the patient has sepsis, there are certain reliable indictors that can shed light on a patient's immune status. These measurable indicators, or *biomarkers*, are reproducible, highly sensitive, and highly specific [117]. Some biomarkers that are indicative of the early, hyper-proinflammatory phase of sepsis are similar to typical markers of inflammation: increased acute-phase reactant protein levels and increased pro-inflammatory cytokine/chemokine levels [117]. According to the Third International Consensus Definition for Sepsis and Septic Shock, elevated lactate is included as a biomarker for sepsis diagnosis [118], as elevations in lactate are correlated with increased mortality in sepsis [119]. Of note, one innate immune system protein called pentraxin 3 (PTX3) was recently assessed for its potential as a biomarker for early sepsis [120–123]. PTX3 is an acute-phase reactant that is highly expressed by phagocytes (such as monocytes/macrophages) following pro-inflammatory responses by the immune system [124]. PTX3 expression was reported in many studies to be upregulated in septic patients compared with non-septic patients, and this increase correlated with clinical severity and unfavorable prognosis [120–126].

Observation shows that most of the accepted sepsis biomarkers are indicators of the early, hyper-proinflammatory stage of sepsis. This likely is because many of the sepsis therapeutics in development focus on blunting the infection-triggered hyper-proinflammatory phase, with medications such as antibiotics, IL-6 inhibitors, and TNF-α inhibitors prevailing. The immunosuppressive stage of sepsis was not previously thought of as a potential area for therapeutic development, likely due to incomplete understanding of the underlying mechanisms leading to immunosuppression in sepsis. With that being said, accurate diagnosis of the late, immunosuppressive stage via biomarkers is critical for the proper administration of therapeutics that could modulate one of the three main underlying sepsis-induced immunosuppressive mechanisms (See Table 2).

7.1. IL-10

Two potential biomarkers of sepsis-induced immunosuppression that were already discussed in this review include decreased lymphocyte numbers in the blood (due to apoptotic mechanisms) and increased numbers of MDSCs in the blood and secondary lymphoid organs. Related to the aspect of immune cell exhaustion discussed earlier, another valuable marker of late sepsis that has been widely accepted is elevated levels of IL-10 in the serum [127]. Not only does the patient's IL-10 level correlate with his/her sepsis severity score, but it also has been reported to be a major risk factor for increased clinical severity [128]. Many immune cells (both innate and adaptive) produce IL-10, as well as epithelial cells and keratinocytes [129]. Physiologically, IL-10 plays an important anti-inflammatory role in immune system homeostasis by balancing the pro-inflammatory response. IL-10 exerts its anti-inflammatory effect by inhibiting the release of pro-inflammatory cytokines, decreasing antigen presentation, and decreasing phagocytosis [129]. While IL-10 is crucial for normal immune cell function, sustained release of IL-10 can lead to immunosuppression by artificially inhibiting the normal functioning of immune cells. This understanding contributes to elevated blood IL-10 levels acting as a possible marker of sepsis-induced immunosuppression.

7.2. GPR18

More recently, the transmembrane protein GPR18 has been identified as a potential biomarker of sepsis-induced immunosuppression. Elevations in GPR18, a G-protein-coupled receptor found on immune cells and most abundantly in the spleen, has only recently been considered a marker of late sepsis [130]. GPR18 is the receptor for a specialized pro-resolving mediator (SPM) called resolvin D2 (RvD2) [130] (SPMs are discussed in detail below). In sepsis patients, elevated GPR18 expression on monocytes is correlated with a worse sepsis severity score [131]. Interestingly, low GPR18 expression on neutrophils is correlated with increased severity and a worse prognosis due to sepsis [132]. Taken together, these data provide evidence that GPR18 expression in different cell types may provide a way of "fine tuning" biomarkers in sepsis.

7.3. HLA-DR

Human leukocyte antigens (HLAs) are a family of genes of the MHC family that code for various proteins involved in the human immune system [133]. HLA-DR is an MHC class II glycoprotein complex expressed on APCs (monocytes/macrophages, dendritic cells, and B cells) [133]. Because HLA-DR is an MHC-II protein, it is required for adaptive immune system activation; therefore, researchers correlate increased expression of HLA-DR on APCs with a well-functioning immune system [3]. Recent research has suggested that decreased HLA-DR expression can be a reliable biomarker for sepsis-induced immunosuppression for many reasons. For example, one study reported that septic patients express 70% less HLA-DR than non-septic patients, and this decrease is associated with an increase in the patient's sequential organ failure assessment (SOFA) score [134]. Similarly, another study found that decreased HLA-DR expression in immune cells within the bone marrow of septic patients was correlated with worse outcomes due to sepsis [135]. Another study found that following stimulation with LPS, immune cells with reduced expression of HLA-DR produce fewer pro-inflammatory cytokines than their immune cell counterparts with higher expression of HLA-DR [70]; this phenomenon is suggestive of immune cell exhaustion. Taken together, the results of these studies suggest that decreased levels of HLA-DR can serve as a potential biomarker of one of the three major mechanisms underlying sepsis-induced immunosuppression (monocyte/macrophage exhaustion).

Table 2. Biomarkers of Sepsis-Induced Immunosuppression.

Biomarker	Role and Correlation with Outcomes Due to Late Sepsis	References
Decreased lymphocyte count	Late-septic patients are reported to have a decreased lymphocyte count compared with non-septic patients. The decrease is thought to be contributed to inappropriate activation of apoptosis pathways. This biomarker is correlated with a higher likelihood of developing secondary infections.	[3,51,89,91,92,96,97]

Table 2. *Cont.*

Biomarker	Role and Correlation with Outcomes Due to Late Sepsis	References
Increased MDSC production/ migration	Late-septic patients have higher levels of MDSCs in the blood and secondary lymphoid organs compared with non-septic patients. While the purpose of increased MDSC production remains controversial, these cells are thought to contribute to worse outcomes due to sepsis due to their immunosuppressive and anti-inflammatory nature.	[3,98,103,104,106–112,136–138]
Increased IL-10	Late-septic patients have elevated levels of IL-10 compared with non-septic patients. Elevated anti-inflammatory responses triggered by IL-10 suggests that the body will not be able to mount an inflammatory response if the patient does develop a secondary infection. This biomarker is correlated with worse severity and worse clinical outcomes due to sepsis.	[3,127,128]
GPR18 expression	Late-septic patients have changes in GPR18 expression on their immune cells compared with non-septic patients. Changes to GPR18 expression suggests that there is decreased specialized pro-resolving mediator (SPM) bioavailability and uncoupled inflammation resolution circuits. Increased GPR18 expression on monocytes and decreased GPR18 expression on neutrophils is correlated with an increase in mortality due to sepsis.	[130–132]
Decreased HLA-DR expression	Late-septic patients have 70% less HLA-DR expression compared with non-septic patients. Decreased HLA-DR expression suggests that the immune system will be unable to activate T-cell-mediated adaptive immune responses if the patient develops a secondary infection. This biomarker is correlated with worse sequential organ failure assessment (SOFA) scores due to sepsis.	[3,70,134,135]

8. Specialized Pro-Resolving Mediators (SPMs) and Immunosuppression in Sepsis

As mentioned earlier, specialized pro-resolving mediators (SPMs) are a group of endogenously produced lipid mediators derived from fatty acids that work to resolve inflammation without being immunosuppressive [15,139–146]. SPMs are enzymatically derived from arachidonic acid (AA), eicosapentaenoic acid (EPA), and docosahexaenoic acid (DHA) [140]. Recent research has shown that SPMs are crucial for actively resolving acute inflammation following infection via interactions with their cognate receptors (found on many cellular targets, including immune cells) [15,140]. SPMs have been shown to resolve inflammation in many ways. One way is by enhancing macrophage functions during inflammation; SPMs have been shown to increase macrophage phagocytic and efferocytic abilities and dampen the macrophage-mediated release of pro-inflammatory cytokines [15,147]. On the other hand, SPMs reduce neutrophil activation and migration [15,148] while increasing their phagocytic ability [127,131]. Overall, they have been shown to mediate inflammation resolution by acting on immune cells (mostly on neutrophils and macrophages) to limit tissue damage and promote healing [140]. Defects in any SPM production pathway can lead to chronic inflammation and can contribute to immunopathology states, further highlighting the importance of endogenous SPM production in maintaining immune system homeostasis [140]. Taken together, SPMs resolve acute inflammation and prevent chronic inflammation.

Resolvins are derived from EPA ("E-series resolvins") and DHA ("D-series resolvins") [148], and their potent pro-resolving activities have been studied in preclinical murine models in the context of early infection or sepsis [144,145,149]. For example, Jundi et al. mapped resolution circuits in sepsis patients for two D-series resolvins, resolvin D1 (RvD1) and resolvin D2 (RvD2) [131]. They found from these studies that there is decreased bioavailability of endogenously produced SPMs present in these patients, ultimately leading to uncoupled inflammation resolution pathways in various immune cells [131]. When these septic patients were treated with ex vivo RvD1 and RvD2, their inflammation resolution pathways normalized [131]. Similarly, Dalli et al. reported that there was a measurable increase in SPM levels in the blood of sepsis non-survivors [150]. Like the conclusions drawn in Jundi et al., the authors suggest this may likely be a reflection of the dissociation of circulating SPMs and their receptors, leading to the uncoupling of inflammation resolution pathways [150].

While there are more studies in the literature that focus on the roles of SPMs in early sepsis, there are very few studies that focus on the role of SPMs in late sepsis. An established and accepted model of late polymicrobial sepsis is the two-hit model of CLP (hit one) followed by *Pseudomonas aeruginosa* secondary infection (hit two) [137,151–153]. This two-hit model enables the observation of how SPMs (administered late in an infection) may impact the host prior to a secondary bacterial lung infection [152,153]. In studies using this murine preclinical model, the administration of RvD2 increased secondary lung bacterial clearance, inflammation, and murine survival after secondary infection [152,153]. While the exact mechanism of this protection is currently unknown, it can be postulated that the SPMs positively enhance the function of certain immune cells such as monocytes, macrophages, and neutrophils in order to help clear the infection without triggering excess inflammation. This possibility is further demonstrated by studies showing that SPM administration lowers the amount of antibiotics needed in order for the host to successfully clear the bacterial infection or bacteria-produced biofilm [142,154–156]. Additionally, D-series resolvins were shown to directly increase macrophage phagocytic functions as a primary mechanism of clearing bacteria in late-sepsis murine models [153] as well as in cystic fibrosis murine models [157,158]. RvD2 was also reported to increase the number of M-MDSCs in CLP mouse spleens compared with sham spleens, and this increase was associated with enhanced bacterial clearance and reduced mortality due to sepsis [152]. These studies provide evidence that SPMs may have therapeutic value in the treatment of sepsis during the immunosuppressive phase.

9. Summary and Conclusions

In summary, three main mechanisms underlying sepsis-induced immunosuppression include lymphocyte apoptosis, monocyte/macrophage exhaustion, and MDSC migration. These mechanisms have one commonality: persistence of the early, hyper-proinflammatory phase of sepsis, which eventually leads to a late, immunosuppressive state. While there have been promising advancements in antibiotic development and preclinical study designs and research techniques, there is still a great need for effective sepsis therapies. Early recognition of late-sepsis biomarkers, such as decreased lymphocyte count, increased MDSC production, increased IL-10 levels, changes in GPR18 expression, and decreased HLA-DR expression, will help with targeted diagnoses and treatments. The use of SPMs is an intriguing possibility to protect the patient from secondary infections and death due to sepsis-induced immunosuppression. Further studies into the therapeutic value of SPMs in the treatment of sepsis-induced immunosuppression is warranted.

Author Contributions: Writing—original draft preparation, C.M.P. and K.Y. Writing—review and editing, C.M.P. and K.Y. All authors have read and agreed to the published version of the manuscript.

Funding: This work was supported by an NIH grant (RO1 AI128202) to K.Y.

Institutional Review Board Statement: Not applicable.

Informed Consent Statement: Not applicable.

Data Availability Statement: No new data were created in this study. Data sharing is not applicable to this article.

Conflicts of Interest: The authors declare no conflicts of interest.

Abbreviations

SPMs	specialized pro-resolving mediators
MDSCs	myeloid-derived suppressor cells
CRS	cytokine release syndrome
ILCs	innate lymphoid cells
PAMPs	pathogen-associated molecular patterns
PRRs	pattern recognition receptors
iPRRs	inhibitory pattern recognition receptors
LPS	lipopolysaccharide
NF-κB	nuclear factor kappa light chain enhancer of activated B cells
APC	antigen-presenting cell
TCR	T-cell receptor
IFN-γ	interferon gamma
TNF-α	tumor necrosis factor alpha
TLR2	Toll-like receptor 2
IL-4	interleukin 4
IL-6	interleukin 6
IL-7	interleukin 7
IL-8	interleukin 8
IL-10	interleukin 10
IL-13	interleukin 13
GM-CSF	granulocyte macrophage colony-stimulating factor
M-CSF	macrophage colony-stimulating factor
ICs	immune complexes
GCs	glucocorticoids
LIF	leukemia inhibitory factor
CTLA-4	cytotoxic T-lymphocyte-associated protein 4
PD-1	programmed cell death protein 1
PD-L1	programmed death ligand 1
CLP	cecal ligation and puncture
HLA-DR	human leukocyte antigen DR isotype
ET	endotoxin tolerance
HMGB1	high mobility group box 1
ROS	reactive oxygen species
Bcl-2	B cell lymphoma 2
FasL	Fas ligand
Bax	Bcl-2-associated X protein
Bak	Bcl-2 antagonist killer 1
LTx	lethal toxin
TUNEL	transferase-mediated dUTP nick-end labeling
G-CSF	granulocyte colony-stimulating factor
CXCL2	chemokine ligand 2
ICU	intensive care unit
M-MDSCs	monocytic MDSCs
PMN-MDSCs	polymorphic MDSCs
GPR18	G-protein-coupled receptor 18
RvD1	resolvin D1
RvD2	resolvin D2
PTX3	pentraxin 3
HLA	human leukocyte antigen
MHC	major histocompatibility complex
APC	antigen-presenting cell
SOFA	sequential organ failure assessment
AA	arachidonic acid
EPA	eicosapentaenoic acid
DHA	docosahexaenoic acid

References

1. Athale, J.; Busch, L.M.; O'Grady, N.P. Cytokine Release Syndrome and Sepsis. *Infect. Dis. Clin. N. Am.* **2022**, *36*, 735–748. [CrossRef] [PubMed]
2. Fajgenbaum, D.C.; June, C.H. Cytokine Storm. *N. Engl. J. Med.* **2020**, *383*, 2255–2273. [CrossRef] [PubMed]
3. Liu, D.; Huang, S.-Y.; Sun, J.-H.; Zhang, H.-C.; Cai, Q.-L.; Gao, C.; Li, L.; Cao, J.; Xu, F.; Zhou, Y.; et al. Sepsis-Induced Immunosuppression: Mechanisms, Diagnosis and Current Treatment Options. *Mil. Med. Res.* **2022**, *9*, 56. [CrossRef] [PubMed]
4. Marshall, J.C. Why Have Clinical Trials in Sepsis Failed? *Trends Mol. Med.* **2014**, *20*, 195–203. [CrossRef] [PubMed]
5. Esposito, R. Tools to Study Adaptive and Innate Immune Response 2021. Available online: https://www.enzolifesciences.com/science-center/technotes/2021/may/tools-to-study-adaptive-and-innate-immune-response/ (accessed on 20 December 2023).
6. Chaplin, D.D. Overview of the Immune Response. *J. Allergy Clin. Immunol.* **2010**, *125*, S3–S23. [CrossRef]
7. Mazzurana, L.; Rao, A.; Van Acker, A.; Mjösberg, J. The Roles for Innate Lymphoid Cells in the Human Immune System. *Semin. Immunopathol.* **2018**, *40*, 407–419. [CrossRef]
8. Janeway, C.J.; Travers, P.; Walport, M. Principles of Innate and Adaptive Immunity. In *Immunobiology: The Immune System in Health and Disease;* Garland Science: New York, NY, USA, 2001.
9. Zigterman, B.G.R.; Dubois, L. Inflammation and Infection: Cellular and Biochemical Processes. *Ned. Tijdschr. Voor Tandheelkd.* **2022**, *129*, 125–129. [CrossRef]
10. Chen, L.; Deng, H.; Cui, H.; Fang, J.; Zuo, Z.; Deng, J.; Li, Y.; Wang, X.; Zhao, L. Inflammatory Responses and Inflammation-Associated Diseases in Organs. *Oncotarget* **2018**, *9*, 7204–7218. [CrossRef]
11. Nedeva, C. Inflammation and Cell Death of the Innate and Adaptive Immune System during Sepsis. *Biomolecules* **2021**, *11*, 1011. [CrossRef]
12. Liu, T.; Zhang, L.; Joo, D.; Sun, S.-C. NF-κB Signaling in Inflammation. *Signal Transduct. Target. Ther.* **2017**, *2*, 17023. [CrossRef]
13. Horwitz, D.A.; Fahmy, T.M.; Piccirillo, C.A.; La Cava, A. Rebalancing Immune Homeostasis to Treat Autoimmune Diseases. *Trends Immunol.* **2019**, *40*, 888–908. [CrossRef] [PubMed]
14. Kourtzelis, I.; Hajishengallis, G.; Chavakis, T. Phagocytosis of Apoptotic Cells in Resolution of Inflammation. *Front. Immunol.* **2020**, *11*, 553. [CrossRef] [PubMed]
15. Serhan, C.N.; Savill, J. Resolution of Inflammation: The Beginning Programs the End. *Nat. Immunol.* **2005**, *6*, 1191–1197. [CrossRef] [PubMed]
16. Rumpret, M.; von Richthofen, H.J.; Peperzak, V.; Meyaard, L. Inhibitory Pattern Recognition Receptors. *J. Exp. Med.* **2022**, *219*, e20211463. [CrossRef] [PubMed]
17. Jendro, M.; Goronzy, J.J.; Weyand, C.M. Structural and Functional Characterization of Hla-Dr Molecules Circulating in the Serum. *Autoimmunity* **1991**, *8*, 289–296. [CrossRef]
18. Cheadle, W.G. The Human Leukocyte Antigens and Their Relationship to Infection. *Am. J. Surg.* **1993**, *165*, 75S–81S. [CrossRef]
19. Petersone, L.; Edner, N.M.; Ovcinnikovs, V.; Heuts, F.; Ross, E.M.; Ntavli, E.; Wang, C.J.; Walker, L.S.K. T Cell/B Cell Collaboration and Autoimmunity: An Intimate Relationship. *Front. Immunol.* **2018**, *9*, 1941. [CrossRef]
20. Xing, Y.; Hogquist, K.A. T-Cell Tolerance: Central and Peripheral. *Cold Spring Harb. Perspect. Biol.* **2012**, *4*, a006957. [CrossRef]
21. Tai, Y.; Wang, Q.; Korner, H.; Zhang, L.; Wei, W. Molecular Mechanisms of T Cells Activation by Dendritic Cells in Autoimmune Diseases. *Front. Pharmacol.* **2018**, *9*, 642. [CrossRef]
22. Sharma, P.; Allison, J.P. The Future of Immune Checkpoint Therapy. *Science* **2015**, *348*, 56–61. [CrossRef]
23. Buchbinder, E.I.; Desai, A. CTLA-4 and PD-1 Pathways: Similarities, Differences, and Implications of Their Inhibition. *Am. J. Clin. Oncol.* **2016**, *39*, 98–106. [CrossRef] [PubMed]
24. Barnig, C.; Bezema, T.; Calder, P.C.; Charloux, A.; Frossard, N.; Garssen, J.; Haworth, O.; Dilevskaya, K.; Levi-Schaffer, F.; Lonsdorfer, E.; et al. Activation of Resolution Pathways to Prevent and Fight Chronic Inflammation: Lessons From Asthma and Inflammatory Bowel Disease. *Front. Immunol.* **2019**, *10*, 1699. [CrossRef] [PubMed]
25. Serhan, C.N.; Clish, C.B.; Brannon, J.; Colgan, S.P.; Chiang, N.; Gronert, K. Novel Functional Sets of Lipid-Derived Mediators with Antiinflammatory Actions Generated from Omega-3 Fatty Acids via Cyclooxygenase 2–Nonsteroidal Antiinflammatory Drugs and Transcellular Processing. *J. Exp. Med.* **2000**, *192*, 1197–1204. [CrossRef] [PubMed]
26. Yang, J.; Zhang, L.; Yu, C.; Yang, X.-F.; Wang, H. Monocyte and Macrophage Differentiation: Circulation Inflammatory Monocyte as Biomarker for Inflammatory Diseases. *Biomark. Res.* **2014**, *2*, 1. [CrossRef]
27. Lam, R.S.; O'Brien-Simpson, N.M.; Holden, J.A.; Lenzo, J.C.; Fong, S.B.; Reynolds, E.C. Unprimed, M1 and M2 Macrophages Differentially Interact with Porphyromonas Gingivalis. *PLoS ONE* **2016**, *11*, e0158629. [CrossRef] [PubMed]
28. Rőszer, T. Understanding the Mysterious M2 Macrophage through Activation Markers and Effector Mechanisms. *Mediat. Inflamm.* **2015**, *2015*, 816460. [CrossRef]
29. Strizova, Z.; Benesova, I.; Bartolini, R.; Novysedlak, R.; Cecrdlova, E.; Foley, L.K.; Striz, I. M1/M2 Macrophages and Their Overlaps—Myth or Reality? *Clin. Sci.* **2023**, *137*, 1067–1093. [CrossRef]
30. Watanabe, S.; Alexander, M.; Misharin, A.V.; Budinger, G.R.S. The Role of Macrophages in the Resolution of Inflammation. *J. Clin. Investig.* **2019**, *129*, 2619–2628. [CrossRef]
31. Sica, A.; Mantovani, A. Macrophage Plasticity and Polarization: In Vivo Veritas. *J. Clin. Investig.* **2012**, *122*, 787–795. [CrossRef]
32. Yao, Y.; Xu, X.-H.; Jin, L. Macrophage Polarization in Physiological and Pathological Pregnancy. *Front. Immunol.* **2019**, *10*, 792. [CrossRef]

33. Zhang, Q.; Sioud, M. Tumor-Associated Macrophage Subsets: Shaping Polarization and Targeting. *Int. J. Mol. Sci.* **2023**, *24*, 7493. [CrossRef] [PubMed]
34. Guilliams, M.; Mildner, A.; Yona, S. Developmental and Functional Heterogeneity of Monocytes. *Immunity* **2018**, *49*, 595–613. [CrossRef] [PubMed]
35. Nahrendorf, M.; Swirski, F.K. Abandoning M1/M2 for a Network Model of Macrophage Function. *Circ. Res.* **2016**, *119*, 414–417. [CrossRef] [PubMed]
36. Xue, J.; Schmidt, S.V.; Sander, J.; Draffehn, A.; Krebs, W.; Quester, I.; De Nardo, D.; Gohel, T.D.; Emde, M.; Schmidleithner, L.; et al. Transcriptome-Based Network Analysis Reveals a Spectrum Model of Human Macrophage Activation. *Immunity* **2014**, *40*, 274–288. [CrossRef] [PubMed]
37. Saas, P.; Chagué, C.; Maraux, M.; Cherrier, T. Toward the Characterization of Human Pro-Resolving Macrophages? *Front. Immunol.* **2020**, *11*, 593300. [CrossRef] [PubMed]
38. Vannella, K.M.; Wynn, T.A. Mechanisms of Organ Injury and Repair by Macrophages. *Annu. Rev. Physiol.* **2017**, *79*, 593–617. [CrossRef] [PubMed]
39. Bah, A.; Vergne, I. Macrophage Autophagy and Bacterial Infections. *Front. Immunol.* **2017**, *8*, 1483. [CrossRef]
40. Cronan, M.R.; Beerman, R.W.; Rosenberg, A.F.; Saelens, J.W.; Johnson, M.G.; Oehlers, S.H.; Sisk, D.M.; Jurcic Smith, K.L.; Medvitz, N.A.; Miller, S.E.; et al. Macrophage Epithelial Reprogramming Underlies Mycobacterial Granuloma Formation and Promotes Infection. *Immunity* **2016**, *45*, 861–876. [CrossRef]
41. Linke, M.; Pham, H.T.T.; Katholnig, K.; Schnöller, T.; Miller, A.; Demel, F.; Schütz, B.; Rosner, M.; Kovacic, B.; Sukhbaatar, N.; et al. Chronic Signaling via the Metabolic Checkpoint Kinase mTORC1 Induces Macrophage Granuloma Formation and Marks Sarcoidosis Progression. *Nat. Immunol.* **2017**, *18*, 293–302. [CrossRef]
42. Zhang, T.; Yu-Jing, L.; Ma, T. Role of Regulation of PD-1 and PD-L1 Expression in Sepsis. *Front. Immunol.* **2023**, *14*, 1029438. [CrossRef]
43. Jiang, X.; Wang, J.; Deng, X.; Xiong, F.; Ge, J.; Xiang, B.; Wu, X.; Ma, J.; Zhou, M.; Li, X.; et al. Role of the Tumor Microenvironment in PD-L1/PD-1-Mediated Tumor Immune Escape. *Mol. Cancer* **2019**, *18*, 10. [CrossRef] [PubMed]
44. Alsaab, H.O.; Sau, S.; Alzhrani, R.; Tatiparti, K.; Bhise, K.; Kashaw, S.K.; Iyer, A.K. PD-1 and PD-L1 Checkpoint Signaling Inhibition for Cancer Immunotherapy: Mechanism, Combinations, and Clinical Outcome. *Front. Pharmacol.* **2017**, *8*, 561. [CrossRef] [PubMed]
45. Karwacz, K.; Bricogne, C.; MacDonald, D.; Arce, F.; Bennett, C.L.; Collins, M.; Escors, D. PD-L1 Co-stimulation Contributes to Ligand-induced T Cell Receptor Down-modulation on CD8 $^{+}$ T Cells. *EMBO Mol. Med.* **2011**, *3*, 581–592. [CrossRef] [PubMed]
46. Song, S.; Yuan, P.; Wu, H.; Chen, J.; Fu, J.; Li, P.; Lu, J.; Wei, W. Dendritic Cells with an Increased PD-L1 by TGF-β Induce T Cell Anergy for the Cytotoxicity of Hepatocellular Carcinoma Cells. *Int. Immunopharmacol.* **2014**, *20*, 117–123. [CrossRef] [PubMed]
47. Zhou, X.; Zhou, Y.; Ding, Q.; Jiao, Z.; Lu, L.; Yang, N.; Ma, Y.; Chou, K.-Y. High Level Expression of B7H1 Molecules by Keratinocytes Suppresses Xeno- and Allo-Reactions by Inducing Type I Regulatory T Cells. *Transplant. Immunol.* **2009**, *21*, 192–197. [CrossRef]
48. Han, Y.; Liu, D.; Li, L. PD-1/PD-L1 Pathway: Current Researches in Cancer. *Am. J. Cancer Res.* **2020**, *10*, 727–742.
49. Pardoll, D.M. The Blockade of Immune Checkpoints in Cancer Immunotherapy. *Nat. Rev. Cancer* **2012**, *12*, 252–264. [CrossRef]
50. Hotchkiss, R.S.; Colston, E.; Yende, S.; Crouser, E.D.; Martin, G.S.; Albertson, T.; Bartz, R.R.; Brakenridge, S.C.; Delano, M.J.; Park, P.K.; et al. Immune Checkpoint Inhibition in Sepsis: A Phase 1b Randomized Study to Evaluate the Safety, Tolerability, Pharmacokinetics, and Pharmacodynamics of Nivolumab. *Intensive Care Med.* **2019**, *45*, 1360–1371. [CrossRef]
51. Yang, L.; Gao, Q.; Li, Q.; Guo, S. PD-L1 Blockade Improves Survival in Sepsis by Reversing Monocyte Dysfunction and Immune Disorder. *Inflammation* **2023**, 1–15. [CrossRef]
52. Ruan, W.-S.; Feng, M.-X.; Xu, J.; Xu, Y.-G.; Song, C.-Y.; Lin, L.-Y.; Li, L.; Lu, Y.-Q. Early Activation of Myeloid-Derived Suppressor Cells Participate in Sepsis-Induced Immune Suppression via PD-L1/PD-1 Axis. *Front. Immunol.* **2020**, *11*, 1299. [CrossRef]
53. Patera, A.C.; Drewry, A.M.; Chang, K.; Beiter, E.R.; Osborne, D.; Hotchkiss, R.S. Frontline Science: Defects in Immune Function in Patients with Sepsis Are Associated with PD-1 or PD-L1 Expression and Can Be Restored by Antibodies Targeting PD-1 or PD-L1. *J. Leukoc. Biol.* **2016**, *100*, 1239–1254. [CrossRef] [PubMed]
54. Wilson, J.K.; Zhao, Y.; Singer, M.; Spencer, J.; Shankar-Hari, M. Lymphocyte Subset Expression and Serum Concentrations of PD-1/PD-L1 in Sepsis—Pilot Study. *Crit. Care* **2018**, *22*, 95. [CrossRef] [PubMed]
55. Zhang, Y.; Li, J.; Lou, J.; Zhou, Y.; Bo, L.; Zhu, J.; Zhu, K.; Wan, X.; Cai, Z.; Deng, X. Upregulation of Programmed Death-1 on T Cells and Programmed Death Ligand-1 on Monocytes in Septic Shock Patients. *Crit. Care* **2011**, *15*, R70. [CrossRef] [PubMed]
56. Biswas, S.K.; Lopez-Collazo, E. Endotoxin Tolerance: New Mechanisms, Molecules and Clinical Significance. *Trends Immunol.* **2009**, *30*, 475–487. [CrossRef] [PubMed]
57. Cavaillon, J.-M.; Adib-Conquy, M. Bench-to-Bedside Review: Endotoxin Tolerance as a Model of Leukocyte Reprogramming in Sepsis. *Crit. Care* **2006**, *10*, 233. [CrossRef] [PubMed]
58. Hoppstädter, J.; Dembek, A.; Linnenberger, R.; Dahlem, C.; Barghash, A.; Fecher-Trost, C.; Fuhrmann, G.; Koch, M.; Kraegeloh, A.; Huwer, H.; et al. Toll-Like Receptor 2 Release by Macrophages: An Anti-Inflammatory Program Induced by Glucocorticoids and Lipopolysaccharide. *Front. Immunol.* **2019**, *10*, 1634. [CrossRef]
59. Pradhan, K.; Yi, Z.; Geng, S.; Li, L. Development of Exhausted Memory Monocytes and Underlying Mechanisms. *Front. Immunol.* **2021**, *12*, 778830. [CrossRef]

60. Cavaillon, J.M.; Adib-Conquy, M.; Cloëz-Tayarani, I.; Fitting, C. Immunodepression in Sepsis and SIRS Assessed by Ex Vivo Cytokine Production Is Not a Generalized Phenomenon: A Review. *J. Endotoxin Res.* **2001**, *7*, 85–93. [CrossRef]
61. Hotchkiss, R.S.; Monneret, G.; Payen, D. Sepsis-Induced Immunosuppression: From Cellular Dysfunctions to Immunotherapy. *Nat. Rev. Immunol.* **2013**, *13*, 862–874. [CrossRef]
62. Bick, A.; Buys, W.; Engler, A.; Madel, R.; Atia, M.; Faro, F.; Westendorf, A.M.; Limmer, A.; Buer, J.; Herbstreit, F.; et al. Immune Hyporeactivity to Bacteria and Multiple TLR-Ligands, yet No Response to Checkpoint Inhibition in Patients Just after Meeting Sepsis-3 Criteria. *PLoS ONE* **2022**, *17*, e0273247. [CrossRef]
63. Boomer, J.S.; To, K.; Chang, K.C.; Takasu, O.; Osborne, D.F.; Walton, A.H.; Bricker, T.L.; Jarman, S.D.; Kreisel, D.; Krupnick, A.S.; et al. Immunosuppression in Patients Who Die of Sepsis and Multiple Organ Failure. *JAMA* **2011**, *306*, 2594. [CrossRef]
64. Blackwell, T.S.; Blackwell, T.R.; Christman, J.W. Induction of Endotoxin Tolerance Depletes Nuclear Factor-κB and Suppresses Its Activation in Rat Alveolar Macrophages. *J. Leukoc. Biol.* **1997**, *62*, 885–891. [CrossRef] [PubMed]
65. Chan, C.; Li, L.; McCall, C.E.; Yoza, B.K. Endotoxin Tolerance Disrupts Chromatin Remodeling and NF-κB Transactivation at the IL-1β Promoter. *J. Immunol.* **2005**, *175*, 461–468. [CrossRef]
66. Del Fresno, C.; García-Rio, F.; Gómez-Piña, V.; Soares-Schanoski, A.; Fernández-Ruíz, I.; Jurado, T.; Kajiji, T.; Shu, C.; Marín, E.; Gutierrez Del Arroyo, A.; et al. Potent Phagocytic Activity with Impaired Antigen Presentation Identifying Lipopolysaccharide-Tolerant Human Monocytes: Demonstration in Isolated Monocytes from Cystic Fibrosis Patients. *J. Immunol.* **2009**, *182*, 6494–6507. [CrossRef] [PubMed]
67. Deng, H.; Maitra, U.; Morris, M.; Li, L. Molecular Mechanism Responsible for the Priming of Macrophage Activation. *J. Biol. Chem.* **2013**, *288*, 3897–3906. [CrossRef]
68. Dobrovolskaia, M.A.; Vogel, S.N. Toll Receptors, CD14, and Macrophage Activation and Deactivation by LPS. *Microbes Infect.* **2002**, *4*, 903–914. [CrossRef] [PubMed]
69. Foster, S.L.; Medzhitov, R. Gene-Specific Control of the TLR-Induced Inflammatory Response. *Clin. Immunol.* **2009**, *130*, 7–15. [CrossRef] [PubMed]
70. Hoogendijk, A.J.; Garcia-Laorden, M.I.; Van Vught, L.A.; Wiewel, M.A.; Belkasim-Bohoudi, H.; Duitman, J.; Horn, J.; Schultz, M.J.; Scicluna, B.P.; Van 'T Veer, C.; et al. Sepsis Patients Display a Reduced Capacity to Activate Nuclear Factor-κB in Multiple Cell Types*. *Crit. Care Med.* **2017**, *45*, e524–e531. [CrossRef]
71. Huang, X.; Venet, F.; Wang, Y.L.; Lepape, A.; Yuan, Z.; Chen, Y.; Swan, R.; Kherouf, H.; Monneret, G.; Chung, C.-S.; et al. PD-1 Expression by Macrophages Plays a Pathologic Role in Altering Microbial Clearance and the Innate Inflammatory Response to Sepsis. *Proc. Natl. Acad. Sci. USA* **2009**, *106*, 6303–6308. [CrossRef] [PubMed]
72. Zhou, M.; Aziz, M.; Yen, H.-T.; Ma, G.; Murao, A.; Wang, P. Extracellular CIRP Dysregulates Macrophage Bacterial Phagocytosis in Sepsis. *Cell Mol. Immunol.* **2022**, *20*, 80–93. [CrossRef]
73. Chen, W.; Qiang, X.; Wang, Y.; Zhu, S.; Li, J.; Babaev, A.; Yang, H.; Gong, J.; Becker, L.; Wang, P.; et al. Identification of Tetranectin-Targeting Monoclonal Antibodies to Treat Potentially Lethal Sepsis. *Sci. Transl. Med.* **2020**, *12*, eaaz3833. [CrossRef] [PubMed]
74. Yang, G.; Zhang, Q.; Tan, J.; Xiong, Y.; Liang, Y.; Yan, J.; Gu, F.; Xu, Y. HMGB1 Induces Macrophage Pyroptosis in Chronic Endometritis. *Int. Immunopharmacol.* **2023**, *123*, 110706. [CrossRef] [PubMed]
75. Zhang, W.; Jiang, H.; Wu, G.; Huang, P.; Wang, H.; An, H.; Liu, S.; Zhang, W. The Pathogenesis and Potential Therapeutic Targets in Sepsis. *MedComm* **2023**, *4*, e418. [CrossRef] [PubMed]
76. Wang, Y.; Zhang, W.; Xu, Y.; Wu, D.; Gao, Z.; Zhou, J.; Qian, H.; He, B.; Wang, G. Extracellular HMGB1 Impairs Macrophage-Mediated Efferocytosis by Suppressing the Rab43-Controlled Cell Surface Transport of CD91. *Front. Immunol.* **2022**, *13*, 767630. [CrossRef] [PubMed]
77. Parker, K.H.; Sinha, P.; Horn, L.A.; Clements, V.K.; Yang, H.; Li, J.; Tracey, K.J.; Ostrand-Rosenberg, S. HMGB1 Enhances Immune Suppression by Facilitating the Differentiation and Suppressive Activity of Myeloid-Derived Suppressor Cells. *Cancer Res.* **2014**, *74*, 5723–5733. [CrossRef] [PubMed]
78. Döcke, W.-D.; Randow, F.; Syrbe, U.; Krausch, D.; Asadullah, K.; Reinke, P.; Volk, H.-D.; Kox, W. Monocyte Deactivation in Septic Patients: Restoration by IFN-γ Treatment. *Nat. Med.* **1997**, *3*, 678–681. [CrossRef] [PubMed]
79. Kim, E.Y.; Ner-Gaon, H.; Varon, J.; Cullen, A.M.; Guo, J.; Choi, J.; Barragan-Bradford, D.; Higuera, A.; Pinilla-Vera, M.; Short, S.A.P.; et al. Post-Sepsis Immunosuppression Depends on NKT Cell Regulation of mTOR/IFN-γ in NK Cells. *J. Clin. Investig.* **2020**, *130*, 3238–3252. [CrossRef] [PubMed]
80. Shen, S.; Shao, Y.; Li, C. Different Types of Cell Death and Their Shift in Shaping Disease. *Cell Death Discov.* **2023**, *9*, 284. [CrossRef]
81. Gao, W.; Wang, X.; Zhou, Y.; Wang, X.; Yu, Y. Autophagy, Ferroptosis, Pyroptosis, and Necroptosis in Tumor Immunotherapy. *Sig Transduct. Target. Ther.* **2022**, *7*, 196. [CrossRef]
82. Bertheloot, D.; Latz, E.; Franklin, B.S. Necroptosis, Pyroptosis and Apoptosis: An Intricate Game of Cell Death. *Cell Mol. Immunol.* **2021**, *18*, 1106–1121. [CrossRef]
83. Tang, D.; Kang, R.; Berghe, T.V.; Vandenabeele, P.; Kroemer, G. The Molecular Machinery of Regulated Cell Death. *Cell Res.* **2019**, *29*, 347–364. [CrossRef]
84. Rathmell, J.C.; Thompson, C.B. Pathways of Apoptosis in Lymphocyte Development, Homeostasis, and Disease. *Cell* **2002**, *109*, S97–S107. [CrossRef] [PubMed]
85. Alberts, B. (Ed.) *Molecular Biology of the Cell*, 4th ed.; Garland: New York, NY, USA, 2002; ISBN 978-0-8153-3218-3.

86. Tsujimoto, Y. Role of Bcl-2 Family Proteins in Apoptosis: Apoptosomes or Mitochondria?: Role of Bcl-2 Family Proteins in Apoptosis. *Genes. Cells* **1998**, *3*, 697–707. [CrossRef]
87. Peña-Blanco, A.; García-Sáez, A.J. Bax, Bak and beyond—Mitochondrial Performance in Apoptosis. *FEBS J.* **2018**, *285*, 416–431. [CrossRef]
88. Sheikh Motahar Vahedi, H.; Bagheri, A.; Jahanshir, A.; Seyedhosseini, J.; Vahidi, E. Association of Lymphopenia with Short Term Outcomes of Sepsis Patients; a Brief Report. *Arch. Acad. Emerg. Med.* **2019**, *7*, e14.
89. Carrero, J.; Unanue, E. Lymphocyte Apoptosis as an Immune Subversion Strategy of Microbial Pathogens. *Trends Immunol.* **2006**, *27*, 497–503. [CrossRef] [PubMed]
90. Van Hauwermeiren, F.; Van Opdenbosch, N.; Van Gorp, H.; De Vasconcelos, N.; Van Loo, G.; Vandenabeele, P.; Kanneganti, T.-D.; Lamkanfi, M. *Bacillus Anthracis* Induces NLRP3 Inflammasome Activation and Caspase-8–Mediated Apoptosis of Macrophages to Promote Lethal Anthrax. *Proc. Natl. Acad. Sci. USA* **2022**, *119*, e2116415119. [CrossRef] [PubMed]
91. Hotchkiss, R.S.; Swanson, P.E.; Knudson, C.M.; Chang, K.C.; Cobb, J.P.; Osborne, D.F.; Zollner, K.M.; Buchman, T.G.; Korsmeyer, S.J.; Karl, I.E. Overexpression of Bcl-2 in Transgenic Mice Decreases Apoptosis and Improves Survival in Sepsis. *J. Immunol.* **1999**, *162*, 4148–4156. [CrossRef]
92. Wang, S.D.; Huang, K.J.; Lin, Y.S.; Lei, H.Y. Sepsis-Induced Apoptosis of the Thymocytes in Mice. *J. Immunol.* **1994**, *152*, 5014–5021. [CrossRef]
93. Luan, Y.; Yao, Y.; Xiao, X.; Sheng, Z. Insights into the Apoptotic Death of Immune Cells in Sepsis. *J. Interferon Cytokine Res.* **2015**, *35*, 17–22. [CrossRef]
94. Luan, Y.-Y.; Dong, N.; Xie, M.; Xiao, X.-Z.; Yao, Y.-M. The Significance and Regulatory Mechanisms of Innate Immune Cells in the Development of Sepsis. *J. Interferon Cytokine Res.* **2014**, *34*, 2–15. [CrossRef] [PubMed]
95. Matsuda, N.; Teramae, H.; Futatsugi, M.; Takano, K.; Yamamoto, S.; Tomita, K.; Suzuki, T.; Yokoo, H.; Koike, K.; Hattori, Y. Up-Regulation of Histamine H 4 Receptors Contributes to Splenic Apoptosis in Septic Mice: Counteraction of the Antiapoptotic Action of Nuclear Factor-κB. *J. Pharmacol. Exp. Ther.* **2010**, *332*, 730–737. [CrossRef] [PubMed]
96. Bo, L.; Wang, F.; Zhu, J.; Li, J.; Deng, X. Granulocyte-Colony Stimulating Factor (G-CSF) and Granulocyte-Macrophage Colony Stimulating Factor (GM-CSF) for Sepsis: A Meta-Analysis. *Crit. Care* **2011**, *15*, R58. [CrossRef] [PubMed]
97. Francois, B.; Jeannet, R.; Daix, T.; Walton, A.H.; Shotwell, M.S.; Unsinger, J.; Monneret, G.; Rimmelé, T.; Blood, T.; Morre, M.; et al. Interleukin-7 Restores Lymphocytes in Septic Shock: The IRIS-7 Randomized Clinical Trial. *JCI Insight* **2018**, *3*, e98960. [CrossRef]
98. Zhang, W.; Fang, X.; Gao, C.; Song, C.; He, Y.; Zhou, T.; Yang, X.; Shang, Y.; Xu, J. MDSCs in Sepsis-Induced Immunosuppression and Its Potential Therapeutic Targets. *Cytokine Growth Factor. Rev.* **2023**, *69*, 90–103. [CrossRef]
99. Veglia, F.; Sanseviero, E.; Gabrilovich, D.I. Myeloid-Derived Suppressor Cells in the Era of Increasing Myeloid Cell Diversity. *Nat. Rev. Immunol.* **2021**, *21*, 485–498. [CrossRef]
100. Goldmann, O.; Beineke, A.; Medina, E. Identification of a Novel Subset of Myeloid-Derived Suppressor Cells During Chronic Staphylococcal Infection That Resembles Immature Eosinophils. *J. Infect. Dis.* **2017**, *216*, 1444–1451. [CrossRef]
101. Bronte, V.; Brandau, S.; Chen, S.-H.; Colombo, M.P.; Frey, A.B.; Greten, T.F.; Mandruzzato, S.; Murray, P.J.; Ochoa, A.; Ostrand-Rosenberg, S.; et al. Recommendations for Myeloid-Derived Suppressor Cell Nomenclature and Characterization Standards. *Nat. Commun.* **2016**, *7*, 12150. [CrossRef]
102. Talmadge, J.E.; Gabrilovich, D.I. History of Myeloid-Derived Suppressor Cells. *Nat. Rev. Cancer* **2013**, *13*, 739–752. [CrossRef]
103. Brudecki, L.; Ferguson, D.A.; McCall, C.E.; El Gazzar, M. Myeloid-Derived Suppressor Cells Evolve during Sepsis and Can Enhance or Attenuate the Systemic Inflammatory Response. *Infect. Immun.* **2012**, *80*, 2026–2034. [CrossRef]
104. Cuenca, A.G.; Delano, M.J.; Kelly-Scumpia, K.M.; Moreno, C.; Scumpia, P.O.; LaFace, D.M.; Heyworth, P.G.; Efron, P.A.; Moldawer, L.L. A Paradoxical Role for Myeloid-Derived Suppressor Cells in Sepsis and Trauma. *Mol. Med.* **2011**, *17*, 281–292. [CrossRef]
105. Delano, M.J.; Scumpia, P.O.; Weinstein, J.S.; Coco, D.; Nagaraj, S.; Kelly-Scumpia, K.M.; O'Malley, K.A.; Wynn, J.L.; Antonenko, S.; Al-Quran, S.Z.; et al. MyD88-Dependent Expansion of an Immature GR-1+CD11b+ Population Induces T Cell Suppression and Th2 Polarization in Sepsis. *J. Exp. Med.* **2007**, *204*, 1463–1474. [CrossRef] [PubMed]
106. Mathias, B.; Delmas, A.L.; Ozrazgat-Baslanti, T.; Vanzant, E.L.; Szpila, B.E.; Mohr, A.M.; Moore, F.A.; Brakenridge, S.C.; Brumback, B.A.; Moldawer, L.L.; et al. Human Myeloid-Derived Suppressor Cells Are Associated with Chronic Immune Suppression After Severe Sepsis/Septic Shock. *Ann. Surg.* **2017**, *265*, 827–834. [CrossRef] [PubMed]
107. Landoni, V.I.; Martire-Greco, D.; Rodriguez-Rodrigues, N.; Chiarella, P.; Schierloh, P.; Isturiz, M.A.; Fernández, G.C. Immature Myeloid Gr-1+ CD11b+ Cells from Lipopolysaccharide-Immunosuppressed Mice Acquire Inhibitory Activity in the Bone Marrow and Migrate to Lymph Nodes to Exert Their Suppressive Function. *Clin. Sci.* **2016**, *130*, 259–271. [CrossRef] [PubMed]
108. Tang, H.; Li, H.; Sun, Z. Targeting Myeloid-Derived Suppressor Cells for Cancer Therapy. *Cancer Biol. Med.* **2021**, *18*, 992–1009. [CrossRef]
109. Uhel, F.; Azzaoui, I.; Grégoire, M.; Pangault, C.; Dulong, J.; Tadié, J.-M.; Gacouin, A.; Camus, C.; Cynober, L.; Fest, T.; et al. Early Expansion of Circulating Granulocytic Myeloid-Derived Suppressor Cells Predicts Development of Nosocomial Infections in Patients with Sepsis. *Am. J. Respir. Crit. Care Med.* **2017**, *196*, 315–327. [CrossRef]
110. Arocena, A.R.; Onofrio, L.I.; Pellegrini, A.V.; Carrera Silva, A.E.; Paroli, A.; Cano, R.C.; Aoki, M.P.; Gea, S. Myeloid-derived Suppressor Cells Are Key Players in the Resolution of Inflammation during a Model of Acute Infection. *Eur. J. Immunol.* **2014**, *44*, 184–194. [CrossRef]

111. Sander, L.E.; Sackett, S.D.; Dierssen, U.; Beraza, N.; Linke, R.P.; Müller, M.; Blander, J.M.; Tacke, F.; Trautwein, C. Hepatic Acute-Phase Proteins Control Innate Immune Responses during Infection by Promoting Myeloid-Derived Suppressor Cell Function. *J. Exp. Med.* **2010**, *207*, 1453–1464. [CrossRef]
112. Schrijver, I.T.; Karakike, E.; Théroude, C.; Baumgartner, P.; Harari, A.; Giamarellos-Bourboulis, E.J.; Calandra, T.; Roger, T. High Levels of Monocytic Myeloid-Derived Suppressor Cells Are Associated with Favorable Outcome in Patients with Pneumonia and Sepsis with Multi-Organ Failure. *Intensive Care Med. Exp.* **2022**, *10*, 5. [CrossRef]
113. Hollen, M.K.; Stortz, J.A.; Darden, D.; Dirain, M.L.; Nacionales, D.C.; Hawkins, R.B.; Cox, M.C.; Lopez, M.-C.; Rincon, J.C.; Ungaro, R.; et al. Myeloid-Derived Suppressor Cell Function and Epigenetic Expression Evolves over Time after Surgical Sepsis. *Crit. Care* **2019**, *23*, 355. [CrossRef]
114. Mayr, F.B.; Yende, S.; Angus, D.C. Epidemiology of Severe Sepsis. *Virulence* **2014**, *5*, 4–11. [CrossRef]
115. Rangel-Frausto, M.S. THE EPIDEMIOLOGY OF BACTERIAL SEPSIS. *Infect. Dis. Clin. North Am.* **1999**, *13*, 299–312. [CrossRef] [PubMed]
116. Tindal, E.W.; Armstead, B.E.; Monaghan, S.F.; Heffernan, D.S.; Ayala, A. Emerging Therapeutic Targets for Sepsis. *Expert. Opin. Ther. Targets* **2021**, *25*, 175–189. [CrossRef] [PubMed]
117. Barichello, T.; Generoso, J.S.; Singer, M.; Dal-Pizzol, F. Biomarkers for Sepsis: More than Just Fever and Leukocytosis—A Narrative Review. *Crit. Care* **2022**, *26*, 14. [CrossRef] [PubMed]
118. Singer, M.; Deutschman, C.S.; Seymour, C.W.; Shankar-Hari, M.; Annane, D.; Bauer, M.; Bellomo, R.; Bernard, G.R.; Chiche, J.-D.; Coopersmith, C.M.; et al. The Third International Consensus Definitions for Sepsis and Septic Shock (Sepsis-3). *JAMA* **2016**, *315*, 801. [CrossRef]
119. Andersen, L.W.; Mackenhauer, J.; Roberts, J.C.; Berg, K.M.; Cocchi, M.N.; Donnino, M.W. Etiology and Therapeutic Approach to Elevated Lactate Levels. *Mayo Clin. Proc.* **2013**, *88*, 1127–1140. [CrossRef]
120. Chen, H.; Li, T.; Yan, S.; Liu, M.; Liu, K.; Zhang, H.; Gao, M.; Xiao, X. Pentraxin-3 Is a Strong Biomarker of Sepsis Severity Identification and Predictor of 90-Day Mortality in Intensive Care Units via Sepsis 3.0 Definitions. *Diagnostics* **2021**, *11*, 1906. [CrossRef]
121. Hamed, S.; Behnes, M.; Pauly, D.; Lepiorz, D.; Barre, M.; Becher, T.; Lang, S.; Akin, I.; Borggrefe, M.; Bertsch, T.; et al. Diagnostic Value of Pentraxin-3 in Patients with Sepsis and Septic Shock in Accordance with Latest Sepsis-3 Definitions. *BMC Infect. Dis.* **2017**, *17*, 554. [CrossRef]
122. Jie, H.; Li, Y.; Pu, X.; Ye, J. Pentraxin 3, a Predicator for 28-Day Mortality in Patients with Septic Shock. *Am. J. Med. Sci.* **2017**, *353*, 242–246. [CrossRef]
123. Lee, Y.T.; Gong, M.; Chau, A.; Wong, W.T.; Bazoukis, G.; Wong, S.H.; Lampropoulos, K.; Xia, Y.; Li, G.; Wong, M.C.S.; et al. Pentraxin-3 as a Marker of Sepsis Severity and Predictor of Mortality Outcomes: A Systematic Review and Meta-Analysis. *J. Infect.* **2018**, *76*, 1–10. [CrossRef]
124. Davoudian, S.; Piovani, D.; Desai, A.; Mapelli, S.N.; Leone, R.; Sironi, M.; Valentino, S.; Silva-Gomes, R.; Stravalaci, M.; Asgari, F.; et al. A Cytokine/PTX3 Prognostic Index as a Predictor of Mortality in Sepsis. *Front. Immunol.* **2022**, *13*, 979232. [CrossRef]
125. Caironi, P.; Masson, S.; Mauri, T.; Bottazzi, B.; Leone, R.; Magnoli, M.; Barlera, S.; Mamprin, F.; Fedele, A.; Mantovani, A.; et al. Pentraxin 3 in Patients with Severe Sepsis or Shock: The ALBIOS Trial. *Eur. J. Clin. Investig.* **2017**, *47*, 73–83. [CrossRef] [PubMed]
126. Wang, G.; Jiang, C.; Fang, J.; Li, Z.; Cai, H. Pentraxin-3 as a Predictive Marker of Mortality in Sepsis: An Updated Systematic Review and Meta-Analysis. *Crit. Care* **2022**, *26*, 167. [CrossRef] [PubMed]
127. Wu, H.-P.; Chen, C.-K.; Chung, K.; Tseng, J.-C.; Hua, C.-C.; Liu, Y.-C.; Chuang, D.-Y.; Yang, C.-H. Serial Cytokine Levels in Patients with Severe Sepsis. *Inflamm. Res.* **2009**, *58*, 385–393. [CrossRef] [PubMed]
128. Gogos, C.A.; Drosou, E.; Bassaris, H.P.; Skoutelis, A. Pro- versus Anti-inflammatory Cytokine Profile in Patients with Severe Sepsis: A Marker for Prognosis and Future Therapeutic Options. *J. Infect. Dis.* **2000**, *181*, 176–180. [CrossRef] [PubMed]
129. Iyer, S.S.; Cheng, G. Role of Interleukin 10 Transcriptional Regulation in Inflammation and Autoimmune Disease. *Crit. Rev. Immunol.* **2012**, *32*, 23–63. [CrossRef] [PubMed]
130. Zhao, M.; Zheng, Z.; Yin, Z.; Zhang, J.; Qin, J.; Wan, J.; Wang, M. Resolvin D2 and Its Receptor GPR18 in Cardiovascular and Metabolic Diseases: A Promising Biomarker and Therapeutic Target. *Pharmacol. Res.* **2023**, *195*, 106832. [CrossRef]
131. Jundi, B.; Lee, D.-H.; Jeon, H.; Duvall, M.G.; Nijmeh, J.; Abdulnour, R.-E.E.; Pinilla-Vera, M.; Baron, R.M.; Han, J.; Voldman, J.; et al. Inflammation Resolution Circuits Are Uncoupled in Acute Sepsis and Correlate with Clinical Severity. *JCI Insight* **2021**, *6*, e148866. [CrossRef]
132. Zhang, L.; Qiu, C.; Yang, L.; Zhang, Z.; Zhang, Q.; Wang, B.; Wang, X. GPR18 Expression on PMNs as Biomarker for Outcome in Patient with Sepsis. *Life Sci.* **2019**, *217*, 49–56. [CrossRef]
133. Wang, E.; Adams, S.; Marincola, F.M.; Stroncek, D.F. Human Leukocyte and Granulocyte Antigens and Antibodies: The HLA and HNA Systems. In *Blood Banking and Transfusion Medicine*; Elsevier: Amsterdam, The Netherlands, 2007; pp. 129–156. ISBN 978-0-443-06981-9.
134. Berry, P.A.; Antoniades, C.G.; Carey, I.; McPhail, M.J.W.; Hussain, M.J.; Davies, E.T.; Wendon, J.A.; Vergani, D. Severity of the Compensatory Anti-Inflammatory Response Determined by Monocyte HLA-DR Expression May Assist Outcome Prediction in Cirrhosis. *Intensive Care Med.* **2011**, *37*, 453–460. [CrossRef]
135. Venet, F.; Monneret, G. Advances in the Understanding and Treatment of Sepsis-Induced Immunosuppression. *Nat. Rev. Nephrol.* **2018**, *14*, 121–137. [CrossRef] [PubMed]

136. Eriksson, E.; Wenthe, J.; Irenaeus, S.; Loskog, A.; Ullenhag, G. Gemcitabine Reduces MDSCs, Tregs and TGFβ-1 While Restoring the Teff/Treg Ratio in Patients with Pancreatic Cancer. *J. Transl. Med.* **2016**, *14*, 282. [CrossRef]
137. Hotchkiss, R.S.; Monneret, G.; Payen, D. Immunosuppression in Sepsis: A Novel Understanding of the Disorder and a New Therapeutic Approach. *Lancet Infect. Dis.* **2013**, *13*, 260–268. [CrossRef]
138. Law, A.M.K.; Valdes-Mora, F.; Gallego-Ortega, D. Myeloid-Derived Suppressor Cells as a Therapeutic Target for Cancer. *Cells* **2020**, *9*, 561. [CrossRef] [PubMed]
139. Abdulnour, R.E.; Sham, H.P.; Douda, D.N.; Colas, R.A.; Dalli, J.; Bai, Y.; Ai, X.; Serhan, C.N.; Levy, B.D. Aspirin-Triggered Resolvin D1 Is Produced during Self-Resolving Gram-Negative Bacterial Pneumonia and Regulates Host Immune Responses for the Resolution of Lung Inflammation. *Mucosal Immunol.* **2016**, *9*, 1278–1287. [CrossRef] [PubMed]
140. Basil, M.C.; Levy, B.D. Specialized Pro-Resolving Mediators: Endogenous Regulators of Infection and Inflammation. *Nat. Rev. Immunol.* **2016**, *16*, 51–67. [CrossRef] [PubMed]
141. Chen, J.; Purvis, G.S.D.; Collotta, D.; Al Zoubi, S.; Sugimoto, M.A.; Cacace, A.; Martin, L.; Colas, R.A.; Collino, M.; Dalli, J.; et al. RvE1 Attenuates Polymicrobial Sepsis-Induced Cardiac Dysfunction and Enhances Bacterial Clearance. *Front. Immunol.* **2020**, *11*, 2080. [CrossRef]
142. Chiang, N.; de la Rosa, X.; Libreros, S.; Serhan, C.N. Novel Resolvin D2 Receptor Axis in Infectious Inflammation. *J. Immunol.* **2017**, *198*, 842–851. [CrossRef]
143. El Kebir, D.; Gjorstrup, P.; Filep, J.G. Resolvin E1 Promotes Phagocytosis-Induced Neutrophil Apoptosis and Accelerates Resolution of Pulmonary Inflammation. *Proc. Natl. Acad. Sci. USA* **2012**, *109*, 14983–14988. [CrossRef]
144. Seki, H.; Fukunaga, K.; Arita, M.; Arai, H.; Nakanishi, H.; Taguchi, R.; Miyasho, T.; Takamiya, R.; Asano, K.; Ishizaka, A.; et al. The Anti-Inflammatory and Proresolving Mediator Resolvin E1 Protects Mice from Bacterial Pneumonia and Acute Lung Injury. *J. Immunol.* **2010**, *184*, 836–843. [CrossRef]
145. Spite, M.; Norling, L.V.; Summers, L.; Yang, R.; Cooper, D.; Petasis, N.A.; Flower, R.J.; Perretti, M.; Serhan, C.N. Resolvin D2 Is a Potent Regulator of Leukocytes and Controls Microbial Sepsis. *Nature* **2009**, *461*, 1287–1291. [CrossRef] [PubMed]
146. Winkler, J.W.; Orr, S.K.; Dalli, J.; Cheng, C.-Y.C.; Sanger, J.M.; Chiang, N.; Petasis, N.A.; Serhan, C.N. Resolvin D4 Stereoassignment and Its Novel Actions in Host Protection and Bacterial Clearance. *Sci. Rep.* **2016**, *6*, 18972. [CrossRef]
147. Gao, J.; Su, Y.; Wang, Z. Lung Inflammation Resolution by RvD1 and RvD2 in a Receptor-Dependent Manner. *Pharmaceutics* **2023**, *15*, 1527. [CrossRef] [PubMed]
148. Serhan, C.N.; Levy, B.D. Resolvins in Inflammation: Emergence of the pro-Resolving Superfamily of Mediators. *J. Clin. Investig.* **2018**, *128*, 2657–2669. [CrossRef] [PubMed]
149. Walker, J.; Dichter, E.; Lacorte, G.; Kerner, D.; Spur, B.; Rodriguez, A.; Yin, K. Lipoxin A4 Increases Survival by Decreasing Systemic Inflammation and Bacterial Load in Sepsis. *Shock* **2011**, *36*, 410–416. [CrossRef]
150. Dalli, J.; Colas, R.A.; Quintana, C.; Barragan-Bradford, D.; Hurwitz, S.; Levy, B.D.; Choi, A.M.; Serhan, C.N.; Baron, R.M. Human Sepsis Eicosanoid and Proresolving Lipid Mediator Temporal Profiles: Correlations with Survival and Clinical Outcomes. *Crit. Care Med.* **2017**, *45*, 58–68. [CrossRef] [PubMed]
151. Muenzer, J.T.; Davis, C.G.; Chang, K.; Schmidt, R.E.; Dunne, W.M.; Coopersmith, C.M.; Hotchkiss, R.S. Characterization and Modulation of the Immunosuppressive Phase of Sepsis. *Infect. Immun.* **2010**, *78*, 1582–1592. [CrossRef] [PubMed]
152. Sundarasivarao, P.Y.K.; Walker, J.M.; Rodriguez, A.; Spur, B.W.; Yin, K. Resolvin D2 Induces Anti-Microbial Mechanisms in a Model of Infectious Peritonitis and Secondary Lung Infection. *Front. Immunol.* **2022**, *13*, 1011944. [CrossRef]
153. Walker, J.M.; Sundarasivarao, P.Y.K.; Thornton, J.M.; Sochacki, K.; Rodriguez, A.; Spur, B.W.; Acharya, N.K.; Yin, K. Resolvin D2 Promotes Host Defense in a 2—Hit Model of Sepsis with Secondary Lung Infection. *Prostaglandins Other Lipid Mediat.* **2022**, *159*, 106617. [CrossRef]
154. Chiang, N.; Fredman, G.; Bäckhed, F.; Oh, S.F.; Vickery, T.; Schmidt, B.A.; Serhan, C.N. Infection Regulates Pro-Resolving Mediators That Lower Antibiotic Requirements. *Nature* **2012**, *484*, 524–528. [CrossRef]
155. Thornton, J.M.; Padovani, C.M.; Rodriguez, A.; Spur, B.W.; Yin, K. Lipoxin A4 Promotes Antibiotic and Monocyte Bacterial Killing in Established Pseudomonas Aeruginosa Biofilm Formed under Hydrodynamic Conditions. *FASEB J.* **2023**, *37*, e23098. [CrossRef] [PubMed]
156. Thornton, J.M.; Walker, J.M.; Sundarasivarao, P.Y.K.; Spur, B.W.; Rodriguez, A.; Yin, K. Lipoxin A4 Promotes Reduction and Antibiotic Efficacy against Pseudomonas Aeruginosa Biofilm. *Prostaglandins Other Lipid Mediat.* **2021**, *152*, 106505. [CrossRef] [PubMed]
157. Codagnone, M.; Cianci, E.; Lamolinara, A.; Mari, V.C.; Nespoli, A.; Isopi, E.; Mattoscio, D.; Arita, M.; Bragonzi, A.; Iezzi, M.; et al. Resolvin D1 Enhances the Resolution of Lung Inflammation Caused by Long-Term Pseudomonas Aeruginosa Infection. *Mucosal Immunol.* **2018**, *11*, 35–49. [CrossRef] [PubMed]
158. Isopi, E.; Mattoscio, D.; Codagnone, M.; Mari, V.C.; Lamolinara, A.; Patruno, S.; D'Aurora, M.; Cianci, E.; Nespoli, A.; Franchi, S.; et al. Resolvin D1 Reduces Lung Infection and Inflammation Activating Resolution in Cystic Fibrosis. *Front. Immunol.* **2020**, *11*, 581. [CrossRef]

biomedicines

MDPI

Article

Cell-Free Nuclear and Mitochondrial DNA as Potential Biomarkers for Assessing Sepsis Severity

Felipe Silva de Miranda [1,2,3,†], Livia Maria A. M. Claudio [4,†], Dayanne Silva M. de Almeida [1,3], Juliana Braga Nunes [1,3], Valério Garrone Barauna [5], Wilson Barros Luiz [1,2,3], Paula Frizzera Vassallo [6] and Luciene Cristina Gastalho Campos [1,2,3,*]

1 Department of Biological Science State, University of Santa Cruz, Ilhéus 45662-900, Bahia, Brazil; felipemiranda2004@hotmail.com (F.S.d.M.); almeida.dsm@gmail.com (D.S.M.d.A.); ju_braga@hotmail.com (J.B.N.); wbluiz@uesc.br (W.B.L.)
2 Postgraduate Program in Biology and Biotechnology of Microorganisms State, University of Santa Cruz, Ilhéus 45662-900, Bahia, Brazil
3 Laboratory of Applied Pathology and Genetics State, University of Santa Cruz, Ilhéus 45662-900, Bahia, Brazil
4 Post Graduation Program in Physiological Sciences, Federal University of Espírito Santo, Vitória 29075-910, Espírito Santo, Brazil; livia.ama@hotmail.com
5 Molecular Physiology Laboratory of Exercise Science, Federal University of Espírito Santo, Vitória 29075-910, Espírito Santo, Brazil; barauna2@gmail.com
6 Clinical Hospital, Federal University of Minas Gerais, Belo Horizonte 30130-100, Minas Gerais, Brazil; pfvassallo1@gmail.com
* Correspondence: lcgcluiz@uesc.br
† These authors contributed equally to this work.

Abstract: Sepsis continues to be a significant public health challenge despite advances in understanding its pathophysiology and management strategies. Therefore, this study evaluated the value of cell-free nuclear DNA (cf-nDNA) and cell-free mitochondrial DNA (cf-mtDNA) for assessing the severity and prognosis of sepsis. Ninety-four patients were divided into three groups: infection (n = 32), sepsis (n = 30), and septic shock (n = 32). Plasma samples were collected at the time of diagnosis, and cfDNA concentrations were determined by qPCR assay. The results showed that plasma cfDNA levels increased with the severity of the disease. To distinguish between patients with infection and those with sepsis, the biomarker $L1PA2_{90}$ achieved the highest AUC of 0.817 (95% CI: 0.725–0.909), demonstrating a sensitivity of 77.0% and a specificity of 79.3%. When cf-nDNA was combined with the SOFA score, there was a significant improvement in the AUC (0.916 (0.853–0.979)), sensitivity (88.1%), and specificity (80.0%). Moreover, patients admitted to the ICU after being diagnosed with sepsis had significantly higher cf-nDNA concentrations. In patients admitted to the ICU, combining cf-nDNA with the SOFA score yielded an AUC of 0.753 (0.622–0.857), with a sensitivity of 95.2% and a specificity of 50.0%. cfDNA can differentiate between patients with infection and those with sepsis. It can also identify patients who are likely to be admitted to the ICU by predicting those with indications for intensive care, suggesting its potential as a biomarker for sepsis.

Keywords: cfDNA; circulating nucleic acids; intensive care unit (ICU); biomarkers; infectious diseases

Citation: de Miranda, F.S.; Claudio, L.M.A.M.; de Almeida, D.S.M.; Nunes, J.B.; Barauna, V.G.; Luiz, W.B.; Vassallo, P.F.; Campos, L.C.G. Cell-Free Nuclear and Mitochondrial DNA as Potential Biomarkers for Assessing Sepsis Severity. *Biomedicines* **2024**, *12*, 933. https://doi.org/10.3390/biomedicines12050933

Academic Editors: Wen-Lin Su, Sheng-Kang Chiu, Chih-Hao Shen, Yi-Ting Chen and Alejandro Rodriguez

Received: 9 January 2024
Revised: 26 March 2024
Accepted: 19 April 2024
Published: 23 April 2024

1. Introduction

Sepsis is a potentially life-threatening condition characterized by organ dysfunction resulting from a dysregulated immune response to infection [1]. It is estimated that 49 million cases of sepsis occur worldwide each year, resulting in 11 million sepsis-related deaths. This condition is considered a severe public health problem [2]. Additionally, it is the primary cause of death in non-cardiac intensive care units (ICUs), with mortality rates that differ based on the socioeconomic characteristics of the country [3].

Early identification and diagnosis of organ dysfunction in patients with infection are directly related to their prognosis [4]. Therefore, treatment should ideally be initiated within the first hour following the diagnosis of sepsis or septic shock. Each hour of delay in treatment could increase the risk of dying from sepsis by up to 8% [5]. Actually, an increase of 2 points in the Sequential Organ Failure Assessment (SOFA) score indicates possible organ dysfunction and is used for sepsis diagnosis. Septic shock is considered if a patient has persisting hypotension requiring vasopressors to maintain a mean arterial pressure (MAP) $\geq$ 65 mm Hg and a serum lactate level > 2 mmol/L, despite adequate volume resuscitation [1,6,7].

Additionally, the most commonly used biomarkers in clinical practice—C-reactive protein (CRP), procalcitonin (PCT), and lactate—have limited sensitivity and specificity for diagnosis and determining severity. The SOFA score has higher sensitivity compared to these biomarkers, but its specificity is limited because various non-infectious conditions can also lead to organ dysfunction [6,8,9]. Therefore, in the absence of a gold standard method [1], it is crucial to assess potential biomarkers that can diagnose sepsis and identify patients with more severe conditions [10]. In this context, molecular biomarkers such as cell-free DNA (cfDNA) are being investigated in sepsis and various other pathologies as potential biomarkers and noninvasive screening tools. These include cancer, trauma, myocardial infarction, stroke, transplant patient monitoring, and prenatal screening for genetic abnormalities [11].

The relationship between cfDNA and various pathologies can be evaluated by examining the molecular characteristics of cfDNA, including plasma concentration, integrity, and cellular origin, among other factors, to determine the severity of these diseases. The significant increase in plasma cfDNA levels may be associated with pathological processes, such as immune and inflammatory responses, as well as organ dysfunction. The integrity of cfDNA is assessed by analyzing fragment sizes, which reflect the underlying cellular pathophysiological conditions. Necrosis is associated with larger fragment sizes, whereas apoptotic release results in shorter fragments. Additionally, the assessment of cell-free mitochondrial DNA (cf-mtDNA) may serve as a more sensitive diagnostic tool compared to cell-free nuclear DNA (cf-nDNA), which is attributed to the lack of nucleosome-associated histone proteins in mitochondria, which leads to the presence of shorter and more abundant fragments in the plasma [11]. All these molecular features, when combined, can serve as assessment tools for septic patients, just as they are currently used in clinical practice in obstetrics and oncology. Early diagnosis and stratification of the severity of this syndrome increase the possibilities for interventions that could have an impact on mortality.

Therefore, this study aimed to assess cell-free nuclear DNA (cf-nDNA) and cell-free mitochondrial DNA (cf-mtDNA) as potential tools for predicting the severity of sepsis. Furthermore, we investigated whether the increased copy numbers of cf-nDNA and cf-mtDNA are associated with ICU admissions in patients with sepsis.

2. Materials and Methods

2.1. Overview

This is an observational, prospective, cross-sectional study conducted at a public university hospital in the capital of Espírito Santo state, Brazil, from March 2017 to December 2018. The study was submitted and approved by the ethical committee of the Federal University of Espírito Santo under the number 2889-718. All research was performed in accordance with the relevant guidelines and regulations of the Brazilian National Health Council (CNS) Resolution 466/2012 [12] and the Declaration of Helsinki [13]. Patients admitted to the ICU, emergency department, and internal medicine were included. Diagnoses of infection, sepsis, and septic shock were made according to Sepsis-3 consensus definitions. The qSOFA and SOFA scores were calculated according to the criteria described by Sepsis-3 [1,6]. Out of a total of one hundred and thirty-one (131) patients enrolled consecutively in the study through convenient nonprobability sampling, ninety-four (94) adult patients diagnosed with infection, sepsis, or septic shock were included, while thirty-seven (37) patients were excluded due to missing data in medical records, such as the absence of laboratory results, SOFA scores, and outcome information (Figure 1). The control group consisted of patients diagnosed with infection, while the case group comprised those diagnosed with sepsis or septic shock. Written informed consent was provided to all patients. Pediatric patients (under 18 years of age), individuals in a palliative care state, immunocompromised individuals, pregnant women, those who declined to provide consent for study participation, and subjects for whom a sepsis diagnosis could not be confirmed were excluded from this study. Blood samples were collected within 30 min after diagnosis and before the initiation of antibiotic treatment. Four milliliters of peripheral blood was collected in BD Vacutainer tubes containing EDTA (BD, Franklin Lakes, NJ, USA) and centrifuged at 2000× *g* for 10 min at room temperature. The plasma was transferred to a microcentrifuge tube and stored at −80 °C for subsequent analysis. The study was reviewed and approved by the Ethics Committee of the Federal University of Espírito Santo under approval number 2889-718. Written informed consent was provided to patients. All research was conducted in compliance with Brazilian human research legislation, specifically the National Health Council (CNS) Resolution 466/2012 [12], and the Declaration of Helsinki [13].

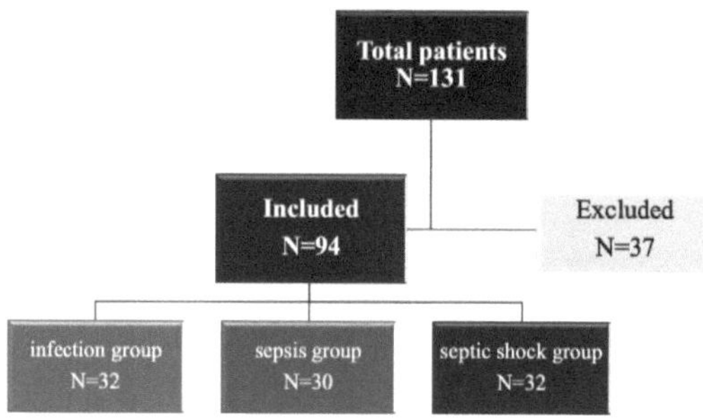

Figure 1. Flow diagram of the patients included in the study.

2.2. *Clinical Data*

Data were collected from the medical records of all patients, including demographic and laboratory data, diagnoses, comorbidities, in-hospital mortality rates, and SOFA and quick SOFA (qSOFA) scores.

2.3. *Analytical Assays*

cfDNA was extracted from plasma by the phenol chloroform method [14]. Briefly, 0.25 mL of plasma in each sample was extracted. The final elution volume was 50 µL. qPCR was performed on QuantStudio™ 3 Real-Time PCR System (Thermo Fisher Scientific, Waltham, MA, USA). For the quantification of cell-free nuclear DNA targeting $L1PA2_{90}$, sequences of the following primer pair were used: Forward (5′-TGCCGCAATAAACATACGTG-3′) and reverse (5′-GACCCAGCCATCCCATTAC-3′); and for targeting $L1PA2_{222}$, sequences of the following primer pair were used: Forward (5′-TGCCGCAATAAACATACGTG-3′) and reverse (5′-AACAACAGGTGCTGGAGAGG-3′) [15]. For the quantification of cell-free mitochondrial DNA, we used a set of primers: Forward (5′-CTATCCGCCATCCCATACATTG-3′) and reverse (5′-ATCGTGTGAGGGTGGGACTG-3′) targeting mitochondrial genome (Mt:15195-15279) with an amplicon of 85 bp [16]. The qPCR conditions were as follows: an initial incubation for 5 min at 95 °C, followed by 40 cycles of 15 s at 95 °C and 1 min at 60 °C. PCR reactions were performed in triplicate using 2 µL of isolated DNA, 5 µL of SYBR Green (Thermo Fisher Scientific, Waltham, MA, USA), 0.1 µL each of forward and reverse primers, and 2.9 µL of DEPC-treated water to reach a final volume of 10 µL. Negative controls (NTCs) were used, each containing 2 µL of DEPC-treated water.

The concentration of cf-nDNA was calculated using the following equation:

$$nDNA = \left(\frac{c}{3.3}\right) * \left(\frac{Velution}{Vplasma}\right)$$

In the equation above, nDNA is the cf-nDNA copy number per milliliter, "c" nDNA concentration (pg/μL) determined by qPCR targeting the nuclear gene L1PA290 or L1PA2222 sequence, and 3.3 pg is the human haploid genome mass. Velution is the volume of cirDNA extract (μL) and Vplasma is the volume of plasma used for the extraction (ml) [17].

Abbreviations: c: cf-nDNA concentration in pg/μL; Velution: elution volume of extracted DNA in μL; Vplasma: volume of plasma used for extraction in mL [17].

cf-mtDNA was calculated based on the following equation:

$$mtDNA = \left(\frac{c * \text{NA}\ (6.02\ *\ 10^{23})}{2 * L * MW}\right) * \left(\frac{Velution}{Vplasma}\right)$$

In the equation above, mtDNA is the cf-mtDNA copy number per milliliter, "c" is the mtDNA mass concentration (g/μL) determined by a qPCR targeting the mitochondrial gene. NA is Avogadro's number (6.02×10^{23} molecules per mole), "L" is the mitochondrial amplicon length (85 pb), and MW is the molecular weight of one nucleotide (g/mol). $V_{elution}$ is the elution volume of cirDNA extract (μL) and V_{plasma} is the volume of plasma used for the extraction (mL) [17].

Abbreviations: c: cf-nDNA concentration in g/μL; Velution: elution volume of extracted DNA in μL; Vplasma: volume of plasma used for extraction in mL [17]. After the equation, the values are converted to a base-2 logarithm. cfDNA quantities are expressed as log2 copy numbers/mL.

2.4. Statistical Analysis

SPSS Statistics version 23 (International Business Machines, New York, NY, USA) and MedCalc version 20.214 (MedCalc Software Ltd., Ostend, Belgium) were used to perform the statistical analyses. A normality test, the Kolmogorov–Smirnov test, was carried out on all quantitative variables. For demographic, clinical, and laboratory characteristics, categorical data were presented as relative frequencies, and quantitative data were presented as medians with interquartile ranges. The data were stratified by diagnosis and compared using Fisher's Exact Test or the Kruskal–Wallis H Test. The results of cfDNA quantification were expressed as log2 copy numbers/mL. The Mann–Whitney U test was used to compare cfDNA levels between the different groups. The Spearman rank correlation coefficient was used to assess the relationship between cfDNA levels and SOFA scores. The predictive performance of cfDNA levels was assessed using ROC analysis. The cut-off point was calculated from the Youden index. The area under the curve (AUC) was compared by the DeLong test. Binary logistic regression was adjusted for comorbidities, age, and sex, showing adjusted odds ratios. A p-value < 0.05 was considered statistically significant.

3. Results

3.1. Patients' Characteristics

A total of 94 patients were included in the study. Of those, 32 patients were diagnosed with infection (34.0%), 30 with sepsis (31.9%), and 32 with septic shock (34.0%). Patients' demographic (Table 1), clinical (Table 2), and laboratory (Table 3) characteristics were evaluated according to their diagnosis.

Table 1. Demographic characteristics according to diagnosis.

	All Cases (N = 94)	Infection (n = 32)	Sepsis (n = 30)	Septic Shock (n = 32)	p-Value †
Age at diagnosis, median (IQR)	**65 (48.7–76)**	54.5 (35–68)	61.5 (49.5–75.2)	75 (68.2–79)	**<0.0001 ***
Male gender (%)	**53.8**	35.5	53.3	71.9	**0.015 ***

Abbreviations: IQR: interquartile range; †: Kruskal–Wallis H test or Fisher's exact test; *: $p < 0.05$.

Table 2. Clinical characteristics according to diagnosis.

	All Cases (N = 94)	Infection (n = 32)	Sepsis (n = 30)	Septic Shock (n = 32)	*p*-Value †
Source of diagnosis, (%)					
Emergency	**33.7**	24.0	36.7	38.7	
Internal medicine	**46.5**	72.0	50.0	22.6	0.397
ICU	**19.8**	4.0	13.3	38.7	
ICU hospitalization (%)	**46.9**	14.8	**54.2**	**70.0**	**<0.0001 ***
Length of stay, median (IQR)	**22 (10–35.5)**	11 (7–23.5)	23.5 (10.2–36.0)	26 (13.5–41)	**0.003 ***
SOFA, median (IQR)	3 (2–8)	1 (0–2.2)	**4 (2–7.2)**	**8 (5.2–10)**	**<0.0001 ***
Positive qSOFA, (%)	54.3	20.0	**53.3**	**87.5**	**<0.0001 ***
Lethality, (%)	**28.0**	9.7	**20.0**	**53.1**	**<0.0001 ***
Comorbidities, (%)					
Hypertension	**45.7**	31.3	33.3	**71.9**	**0.001 ***
Diabetes Mellitus	**24.5**	12.5	26.7	34.4	0.115
Cardiopathy	**28.7**	18.8	16.7	**50.0**	**0.007 ***
Nephropathy	12.8	18.8	3.3	15.6	0.164
Hepatopathy	7.4	3.1	13.3	6.3	0.293
Neoplasia	7.4	3.1	13.3	6.3	0.293
Others	21.3	18.8	16.7	28.1	0.542
None	24.5	**37.5**	30.0	6.3	**0.006 ***
Nosocomial infection, (%)	**60.5**	57.7	48.0	73.3	0.156
Focus of infection, (%)					
Unknown	**31.9**	31.3	26.7	37.5	0.673
Abdominal	**14.9**	12.5	16.7	15.6	0.936
Respiratory	**26.6**	15.6	36.7	28.1	0.168
Genitourinary tract	**17.0**	28.1	10.0	12.5	0.158
Skin-soft tissue/bone-joint	6.4	9.4	6.7	3.1	0.687
Catheter	3.2	3.1	3.3	3.1	1.000

Abbreviations: IQR: interquartile range; ICU: intensive care unit; SOFA: Sequential Organ Failure Assessment; qSOFA: quick Sequential Organ Failure Assessment; †: Kruskal–Wallis H test or Fisher's exact test; *: $p < 0.05$.

Table 3. Laboratory tests according to diagnosis.

	All Cases (N = 94)	Infection (n = 32)	Sepsis (n = 30)	Septic Shock (n = 32)	*p*-Value †
median (IQR)					
Leukocyte (WBCs/mm^3)	12,780 (8520–17,830)	11,270 (9040–16,010)	14,235 (6647–19,995)	13,060 (7464–21,060)	0.912
Platelet ($\times 10^3$/mm^3)	224 (130–300)	244 (154–312)	248 (131–409)	182 (108.5–259)	0.184
INR	1.18 (1.03–1.35)	1.08 (1.00–1.21)	**1.18 (1.03–1.40)**	**1.25 (1.13–1.53)**	**0.014 ***
Total bilirubin (mg/dL)	0.90 (0.52–1.79)	0.58 (0.36–1.05)	**0.97 (0.49–5.02)**	**1.11 (0.69–1.82)**	**0.023 ***
C-reactive protein (mg/L)	112.5 (64.4–128.8)	118.1 (50.1–132.0)	111.6 (55.2–127.3)	110.3 (66.1–129.1)	0.972
Creatinine (mg/dL)	1.49 (0.74–2.32)	0.78 (0.63–1.39)	**1.5 (0.78–2.45)**	**1.83 (1.07–3.11)**	**0.012 ***
Lactate (mmol/L)	1.9 (1.2–2.9)	1.6 (1.1–2.5)	**1.7 (1.1–3.6)**	**2.45 (1.8–3.5)**	**0.030 ***

Abbreviations: IQR: interquartile range; INR: international normalized ratio; †: Kruskal–Wallis H test or Fisher's exact test; *: $p < 0.05$.

The median age was 65 (48.7–76) years. Most patients were male (53.8%) (Table 1). Most diagnoses were from the internal medicine department (46.5%), followed by the emergency department (33.7%) and the ICU (19.8%). A total of 46.9% of the patients were admitted to the ICU, with the majority being diagnosed with septic shock (70.0%) and sepsis (54.2%) ($p < 0.0001$). The median length of hospital stay was 22 (10–35.5) days. The fatality rate was 28%. As expected, patients with prolonged hospital stays have more severe clinical conditions ($p = 0.003$). Patients with sepsis/septic shock also had higher mortality rates, higher SOFA scores, and higher qSOFA positivity compared to the infection group ($p < 0.0001$). The most common comorbidities identified were hypertension (45.7%), heart disease (28.7%), and diabetes (24.5%). The patients in the septic shock group had a higher incidence of hypertension (71.9%; $p = 0.001$) and cardiopathy (50.0%). In contrast, the group with infections had the highest proportion of patients without comorbidities (37.5%; $p = 0.006$). Healthcare-associated infections accounted for 60.5% of the cases. The most common source of infection was unidentified (31.9%), followed by respiratory (26.6%), genitourinary tract (17.0%), and abdominal infections (14.9%) (Table 2).

Regarding laboratory tests, they were collected within 30 min after the diagnosis was made. There were significant differences in the measurements of INR ($p = 0.014$), total bilirubin ($p = 0.023$), creatinine ($p = 0.012$), and lactate ($p = 0.030$), with levels increasing progressively with the severity of the patient's condition (Table 3).

3.2. *cfDNA Concentration according to Sepsis Severity*

Plasma levels of cf-nDNA (L1PA2$_{90}$ and L1PA2$_{222}$) and cf-mtDNA were measured and compared among patients with infection, sepsis, and septic shock. The measurements are expressed as log2 values of the number of copies/mL. The results show higher plasma cfDNA levels correlating with disease severity (Figure 2).

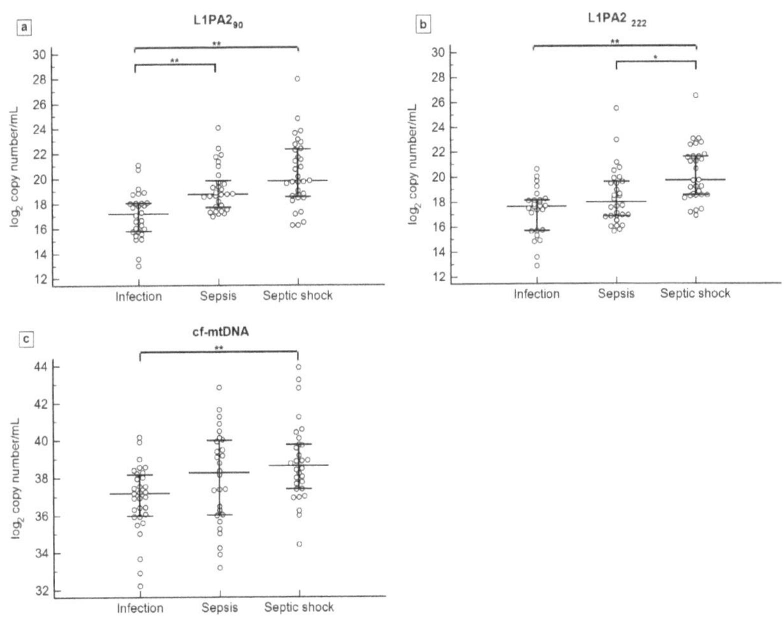

Figure 2. Plasma levels of L1PA2$_{90}$ (**a**), L1PA2$_{222}$ (**b**), and cf-mtDNA (**c**) in patients with infection, sepsis, and septic shock. Data are graphically represented as median and interquartile range (IQR). Abbreviation: * $p < 0.05$, ** $p < 0.0001$.

The median L1PA2$_{90}$ concentration level was 17.28 (15.78–18.05) in patients with infection, 18.75 (17.63–19.96) in patients with sepsis, and 19.77 (18.48–22.33) in patients with septic shock. There are significant differences between patients with infection and those with sepsis ($p < 0.0001$), as well as between patients with infection and those with septic shock ($p < 0.0001$). Median levels of L1PA2$_{222}$ were 17.60 (15.68–18.11) in the infection group, 17.93 (16.83–19.66) in the sepsis group, and 19.69 (18.52–21.64) in the septic shock group. Significant differences were observed between the infection and shock groups ($p < 0.0001$), as well as between the sepsis and septic shock groups ($p = 0.001$). Lastly, cf-mtDNA levels were measured at 37.08 (35.94–38.18) in patients with infection, 38.27 (36.00–40.01) in patients with sepsis, and 38.62 (37.39–39.85) in patients with septic shock.

Significant differences were observed only between patients with infection and those with septic shock ($p < 0.0001$).

3.3. Correlation between cfDNA Levels and SOFA Scores

A moderate, positive, and significant correlation was observed between L1PA2$_{90}$ and the SOFA score (rho = 0.577, 95% CI 0.418–0.702, $p < 0.0001$). In contrast, a weak, positive, and significant correlation was found between L1PA2$_{222}$ and cf-mtDNA (rho = 0.454, 95% CI 0.268–0.608, $p < 0.0001$), and between cf-mtDNA and the SOFA score (rho = 0.267, 95% CI 0.0605–0.451, $p = 0.0121$) (Figure 3).

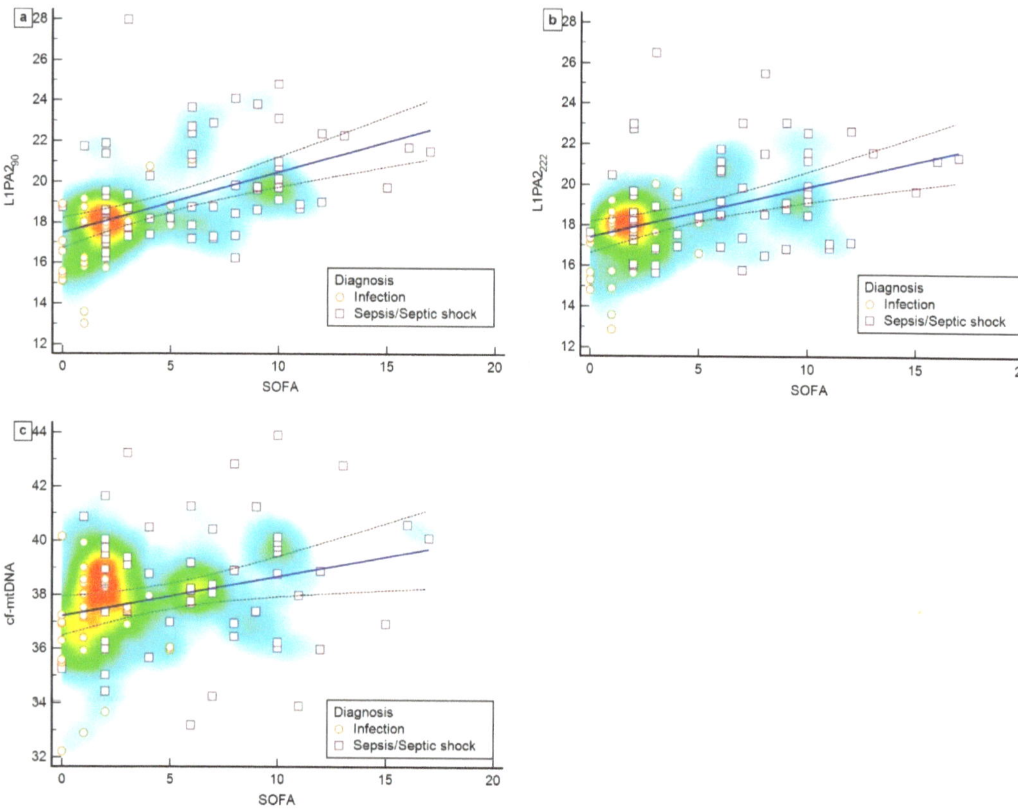

Figure 3. Scatter plot showing the correlation between SOFA score and L1PA2$_{90}$ (**a**), L1PA2$_{222}$ (**b**), and cf-mtDNA (**c**). The orange circle represents patients with infections, while the purple square represents patients with sepsis or septic shock. The colors on the heat map range from blue to red. The higher the density of points, the closer the color approaches red, suggesting cluster formation.

3.4. Potential Diagnostic Value of cfDNA

ROC curves were generated to assess the potential of each cfDNA biomarker to distinguish between patients with infection and those with sepsis or septic shock. The results are shown in Figure 4. The ROC curve analysis indicated that L1PA2$_{90}$ achieved the highest AUC, which was 0.817 (95% CI 0.725–0.909), with a sensitivity of 77.0% and a specificity of 79.3% at a cut-off point of 18.07. Followed by L1PA2$_{222}$, which had an AUC of 0.741 (95% CI: 0.634–0.849), with a sensitivity of 66.6% and a specificity of 81.4% at a cut-off point of 18.27 (Table 4). Additionally, binary logistic regression analysis indicates that the concentration of L1PA2$_{90}$ (OR: 2.067 (95% CI 1.449–2.946); $p < 0.0001$), L1PA2$_{222}$ (OR 1.655 (95% CI 1.240–2.208) $p = 0.001$), and cf-mtDNA (OR 1.415 (95% CI 1.122–1.784);

$p = 0.003$) were significantly associated with disease severity. When cf-nDNA was combined with the SOFA score, there was a significant improvement in AUC (0.916 [95% CI 0.853–0.979]), sensitivity (88.1%), and specificity (80.0%) compared to previously described results ($p < 0.05$).

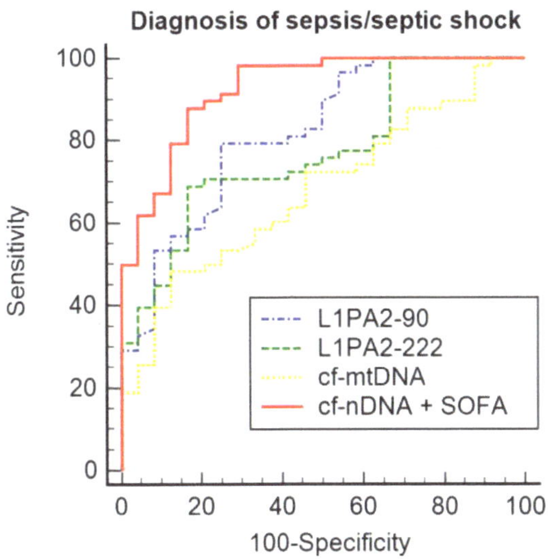

Figure 4. Comparison of ROC curves according to diagnosis.

Table 4. Diagnostic performance of cfDNA for assessing the severity of sepsis using ROC curve and AUC analysis in patients with infection, sepsis, and septic shock.

cfDNA	Cut-Off Value **	AUC (95% CI)	Sensitivity (%)	Specificity (%)
$L1PA2_{90}$	18.07	0.817 (0.725–0.909) *	77.0	79.3
$L1PA2_{222}$	18.27	0.741 (0.634–0.849) *	66.6	81.4
cf-mtDNA	38.58	0.703 (0.596–0.810) *	48.2	90.6
cf-nDNA + SOFA	-	0.916 (0.853–0.979) *	88.1	80.0

Abbreviations: ** cut-off values expressed in log2 copy numbers/mL. Abbreviations: ROC: receiver operating characteristic; AUC: area under the curve; 95% CI: confidence interval at the level 95%; * $p < 0.001$.

3.5. *cfDNA and ICU Hospitalization*

An exploratory analysis was conducted on patients diagnosed in the emergency and internal medicine departments to assess the association between cfDNA levels and ICU admissions. Therefore, 17 patients who were diagnosed directly in the ICU were excluded from this analysis. Patients admitted to the ICU after diagnosis had significantly higher cf-nDNA concentrations than those who remained hospitalized in internal medicine or were discharged from the hospital. This result is not observed with cf-mtDNA, as there are no significant differences in distribution between the groups. The $L1PA2_{90}$ concentrations are significantly higher in the group of patients admitted to the ICU, with levels at 19.12 (17.94–21.53), compared to those not admitted, whose levels were 18.02 (16.42–19.44) ($p = 0.013$). The same pattern is observed for $L1PA2_{222}$, with concentrations of 18.47 (16.91–21.48) in patients admitted to the ICU and 17.52 (15.98–19.07) in those not admitted ($p = 0.049$). However, cf-mtDNA levels showed no significant difference between patients admitted to the ICU, with levels at 38.02 (36.29–40.19), and those who were not hospitalized, with levels at 37.54 (35.80–38.82) ($p = 0.190$) (Figure 5).

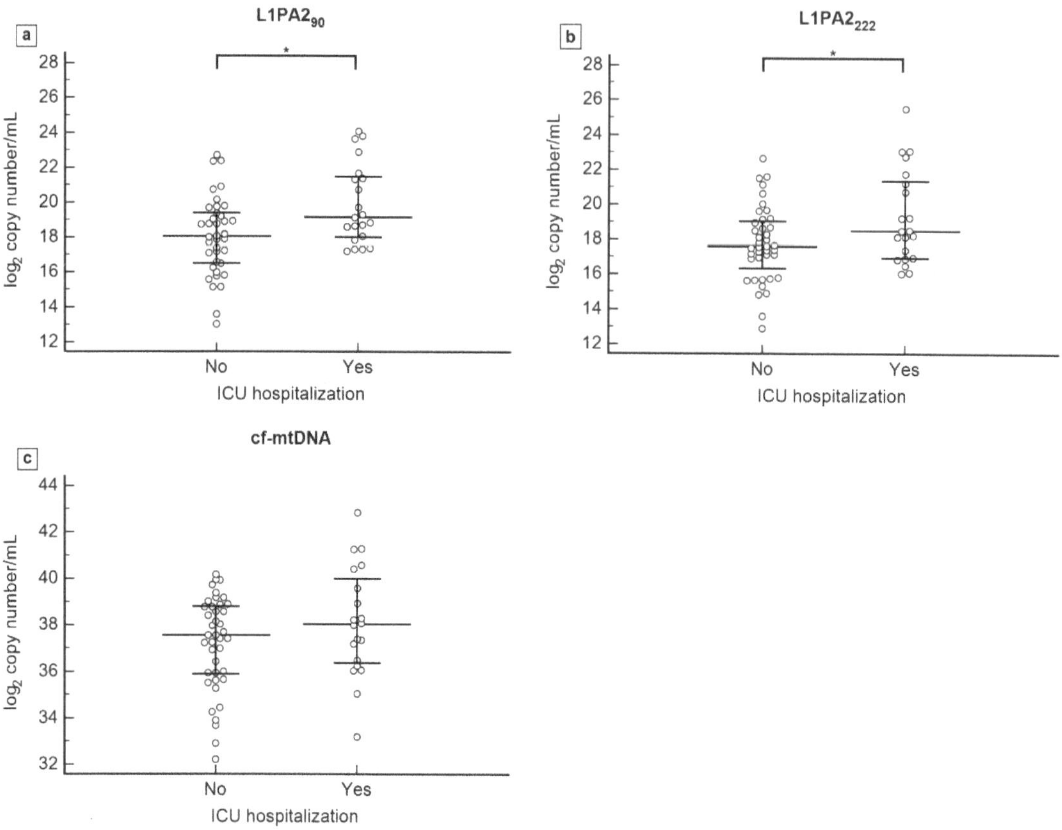

Figure 5. Plasma levels of L1PA2$_{90}$ (**a**), L1PA2$_{222}$ (**b**), and cf-mtDNA (**c**) according to ICU admission. Data are graphically represented as median and interquartile range (IQR). Abbreviation: * $p < 0.05$.

ROC curve analysis of L1PA2$_{90}$ (AUC = 0.692, 95% CI: 0.559–0.806) and L1PA2$_{222}$ (AUC = 0.653, 95% CI: 0.520–0.770) obtained the best results in differentiating patients who were admitted to ICU, with a sensitivity of 100.0% and 66.6% and specificity of 34.2% and 62.5%, respectively. The optimal cut-off point used was 17.17 for L1PA2$_{90}$ and 18.11 for L1PA2$_{222}$. cf-nDNA with SOFA score combined obtained an AUC of 0.753 (95% CI 0.622–0.857), with a sensitivity of 95.2% and a specificity of 50.0% (Table 5). Once again, the combination of SOFA score with cf-nDNA improved accuracy (Figure 6).

Table 5. Performance of cfDNA in predicting transfer to ICU using ROC and AUC analysis in patients with infection, sepsis, and septic shock.

cfDNA	Cut-Off Value **	AUC (95% CI)	Sensitivity z (%)	Specificity (%)
L1PA2$_{90}$	17.17	0.692 (0.559–0.806) *	100.0	34.2
L1PA2$_{222}$	18.11	0.653 (0.520–0.770) *	66.6	62.5
cf-mtDNA	40.17	0.596 (0.463–0.720)	25.0	100.0
cf-nDNA + SOFA	-	0.752 (0.622–0.855) *	95.2	50.0

Abbreviations: ** cut-off values expressed in log2 copy numbers/mL. Abbreviations: ROC: receiver operating characteristic; AUC: area under the curve; 95% CI: confidence interval at the level 95%; * $p < 0.001$.

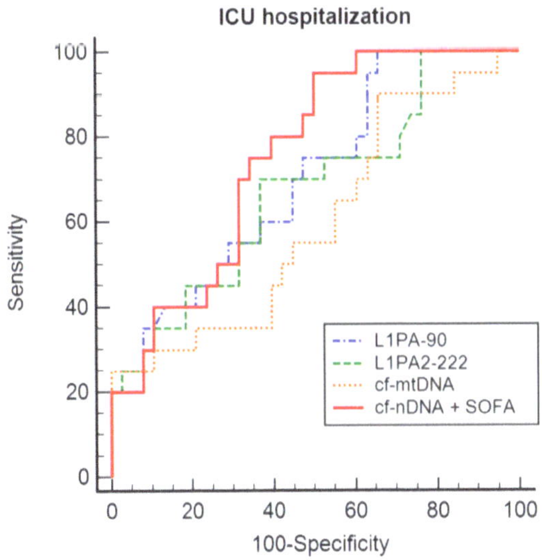

Figure 6. Comparison of ROC curves according to ICU admission.

4. Discussion

Our data indicated that cfDNA concentrations were higher by disease severity, and L1PA2$_{90}$ was able to significantly differentiate between patients with infection, sepsis, and septic shock. ROC curve analysis of cf-nDNA demonstrated improved diagnostic accuracy, and combining cf-nDNA with the SOFA score further increased this accuracy. Similarly, in a subgroup analysis, patients who were hospitalized in the ICU had higher cf-nDNA concentrations.

Sepsis biomarkers have been the subject of intense research [18]. Pierrakos and Vincent estimated that at least 178 biomarkers of sepsis had been reported in the literature [19]. These include acute phase proteins such as C-reactive protein, procalcitonin, pro- and anti-inflammatory cytokines; damage-associated molecular pattern (DAMPs) (e.g., calprotectin and NGAL); endothelial cells and blood–brain barrier markers (e.g., occluding and claudin-5); miRNA (e.g., miR-125a, miR-21); hormones and peptide precursors (e.g., Adrenomedullin, NT-proBNP); and many others [18,20]. Identifying clinically relevant biomarkers would facilitate a more accurate determination of the immune and inflammatory status of a septic patient and could significantly contribute to the stratification of patients who may benefit from a certain therapeutic strategy.

There are several studies evaluating the role of cfDNA in critically ill patients, including those with sepsis. Some of these studies have focused on quantifying cfDNA levels to distinguish septic patients compared to healthy controls or to assess whether the marker is associated with mortality. There are many origins for plasma cfDNA, which can be derived from physiological and pathological processes such as necrosis, apoptosis, exogenous sources, and active cell release, among others. Its origin is heterogeneous, stemming from various organs and cell types. Thus, changes in plasma cfDNA concentrations may reflect the presence of a pathological disorder and its characteristics [11]. Cell-free nuclear DNA and cell-free mitochondrial DNA constitute the plasma cfDNA and have distinct characteristics that need to be differentiated between analyses [21]. The average mean size of cf-nDNA is approximately 166 bp, and larger fragments are associated with necrosis [11]. The analyzed regions are derived from LINE-1 sequences that are distributed across all chromosomes and represent approximately 17% of the entire human genome. Its amplification increases the sensitivity of cfDNA measurement. In turn, cf-mtDNA is more fragmented than cf-nDNA, its mean plasma concentrations are higher, and its increase is associated with pro-inflammatory cytokines such as TNF-α and IL-6 [22,23]. Due to its characteristics and origins, cfDNA can indirectly identify the presence of organ dysfunction in patients with infections. Moreover, using more than one cfDNA target can increase biomarker sensitivity and accuracy.

Our results indicated that patients with sepsis and septic shock have higher levels of circulating cfDNA than patients with infections likely due to organ dysfunction. This was particularly evident with $L1PA2_{90}$, which showed significant differences between the infection group and the group of patients with sepsis and septic shock ($p < 0.0001$). Furthermore, it had the highest AUC of 0.817 (95% CI 0.725–0.909).

Clementi et al. evaluated 27 patients with sepsis and septic shock admitted to the ICU and demonstrated that cf-nDNA can serve as a prognostic marker of severity and has the potential to discriminate between septic and non-septic patients. Additionally, higher levels of cfDNA were associated with acute renal failure requiring renal replacement therapy and an increased duration of mechanical ventilation [24]. Duplessis et al. identified a trend in cf-nDNA concentrations, with higher cfDNA concentrations correlating with increased sepsis severity, and an AUC of 0.650 (95% CI 0.440–0.850). However, they found no significant difference between survivors and non-survivors [25]. Rannikko et al., when evaluating 481 patients with infection, found that the combination of qSOFA score with cf-nDNA predicted 7-day mortality, identifying which patients were at high risk of death. The combination of markers enabled the improvement of AUC from 0.730 (95% CI 0.65–0.82) to 0.770 (95% CI 0.68–0.86) [26].

Regarding cf-mtDNA, our results demonstrate smaller differences between the comparison groups and a lower AUC compared to cf-nDNA, despite having higher specificity when compared to other markers, including cf-nDNA combined with SOFA. Timmermans et al. conducted a prospective observational study in which they collected serial blood samples from 121 patients diagnosed with septic shock, up to the 28th day of their hospitalization. They found that cf-nDNA, cf-mtDNA, and cytokine levels were significantly elevated on the day of diagnosis compared to healthy controls and that these levels were maintained throughout the analyzed period. cf-mtDNA concentrations correlated with IL-1RA, noradrenaline dose, mean arterial pressure, and white blood cell count [27]. Schäfer et al. also found significant differences in cf-mtDNA concentrations between patients diagnosed with sepsis and healthy individuals, with a 123-fold increase [28]. Bhagirath et al., in a translational study evaluating the plasma of 12 septic patients admitted to the ICU, identified concentrations of cf-mtDNA that were 50-fold greater than those in healthy controls and 200-fold greater for cf-nDNA. Additionally, it has been found that a high concentration of cf-mtDNA increases neutrophil viability and activates coagulation and platelets, thereby contributing to the pathophysiology of sepsis [29].

Even with many published studies, there is still a knowledge gap in the link between cfDNA and sepsis. Our study was unprecedented in that it was conducted across the department of internal medicine, emergency, and ICU, not just in intensive care, thereby encompassing a more complete patient profile. Moreover, this study not only encompasses a comprehensive range of diagnoses, including the distinction between infection, sepsis, and septic shock, but also sets itself apart from other studies that typically compare septic patients with healthy controls or non-septic cases exhibiting systemic inflammatory response syndrome (SIRS) in the ICU [10]. cfDNA has the potential to be an efficient biomarker for sepsis and may also assist in differentiating between the stages of the disease. Early identification of organ dysfunction in patients with infection is one of the most important steps in sepsis diagnosis. Thus, one important aspect of this research lies in distinguishing between patients with infection and those with sepsis, rather than employing a control group comprising healthy individuals. This differentiation is crucial and holds relevance in clinical practice, offering valuable insights that can inform targeted interventions and enhance the precision of patient care.

We found that cf-nDNA serves as an independent predictor of severity in patients with infections. Consequently, we conducted an exploratory subanalysis to evaluate the association with ICU admission. Although the analysis is based on a small sample, the analysis seems to confirm a relationship between higher cfDNA levels and increased disease severity. Changes in these biomarkers may serve as valuable indicators, facilitating the identification of hospitalized patients who are at a heightened risk and exhibit a more pronounced need for intensive care. Further research in this context is needed. Despite advances in the study of cfDNA as a biomarker, difficulties in standardizing their quantification remain an obstacle to comparing findings between studies and implementing them in clinical practice [30].

One of the limitations present in total cfDNA quantification analysis is that it does not reveal which tissue these DNA fragments come from. Assessments of cfDNA methylation patterns can be used to identify the tissue origin of cfDNA. This restriction is circumvented by choosing appropriate target genes and their epigenetic signature. This target selection is one of the main factors in achieving relevant and accurate clinical significance [31]. This assessment can be extended to assess specific organ dysfunction in sepsis, such as kidney, liver, or heart damage. The greatest challenge in managing sepsis lies in its varied clinical presentations, which complicate the clinical evaluation and treatment of affected patients. Nevertheless, these challenges can be overcome through the utilization of artificial intelligence tools, specifically with the integration of machine learning with clinical data and molecular markers, such as cfDNA. This approach enables the development of clinical decision support tools pertaining to sepsis and septic shock, ultimately enhancing outcomes for clinicians and facilitating real-time optimization of medical resources [32,33].

5. Conclusions

In conclusion, our data showed that cfDNA concentrations increase with the severity of the clinical state in sepsis. These trends are more pronounced when comparing cf-nDNA with cf-mtDNA. cf-nDNA can accurately differentiate between patients with infection and those with sepsis or septic shock. There is an increase in accuracy, sensitivity, and specificity with the combination of cf-nDNA and organ dysfunction score, which can assist in sepsis diagnosis. Furthermore, the same occurs when assessing ICU hospitalization. The elevated cf-nDNA levels observed in patients admitted to the ICU following a sepsis diagnosis not only corroborate the severity of their clinical condition but also indicate the potential for cfDNA to serve as an early indicator of the need for escalated care. Overall, cfDNA could assist healthcare professionals in assessing the severity of illness, and prognosis, and in making decisions regarding patients' admission to the intensive care unit, thereby showing potential clinical applicability.

Author Contributions: Conception and design: V.G.B., W.B.L., J.B.N., P.F.V. and L.C.G.C.; Assay: F.S.d.M., L.M.A.M.C., J.B.N., D.S.M.d.A. and L.C.G.C.; Data analysis: F.S.d.M., L.M.A.M.C., J.B.N. and D.S.M.d.A.; Writing the article: F.S.d.M., W.B.L. and L.C.G.C.; Critical revision of the article: F.S.d.M., W.B.L. and L.C.G.C.; Obtained funding: V.G.B., W.B.L., P.F.V. and L.C.G.C.; Overall responsibility: L.C.G.C. All authors have read and agreed to the published version of the manuscript.

Funding: L.C.G.C. is supported by Universidade Estadual de Santa Cruz (UESC) (Grant N° 073.11012.2019.0016578-57 and Grant N° 073.11012.2020.0007594-29), FAPESB SUS0046/2018 and W.B.L. FAPESB SUS0027/2021. This study was financed in part by the Coordenação de Aperfeiçoamento de Pessoal de Nível Superior–Brasil (CAPES)–Finance Code 001.

Institutional Review Board Statement: The study was conducted in accordance with the Declaration of Helsinki, and approved by the ethical committee of the Federal University of Espírito Santo under the number 2889-718.

Informed Consent Statement: Informed consent was obtained from all subjects involved in the study.

Data Availability Statement: The datasets used and/or analyzed during the current study are available from the corresponding author on reasonable request.

Acknowledgments: The authors would like to thank the State University of Santa Cruz (UESC) Grants, FAPESB, and Coordenação de Aperfeiçoamento de Pessoal de Nível Superior–Brasil (CAPES).

Conflicts of Interest: The authors declare no conflicts of interest.

References

1. Singer, M.; Deutschman, C.S.; Seymour, C.; Shankar-Hari, M.; Annane, D.; Bauer, M.; Bellomo, R.; Bernard, G.R.; Chiche, J.D.; Coopersmith, C.M.; et al. The Third International Consensus Definitions for Sepsis and Septic Shock (Sepsis-3). *JAMA J. Am. Med. Assoc.* **2016**, *315*, 801–810. [CrossRef] [PubMed]
2. Rudd, K.E.; Johnson, S.C.; Agesa, K.M.; Shackelford, K.A.; Tsoi, D.; Kievlan, D.R.; Colombara, D.V.; Ikuta, K.S.; Kissoon, N.; Finfer, S.; et al. Global, Regional, and National Sepsis Incidence and Mortality, 1990–2017: Analysis for the Global Burden of Disease Study. *Lancet* **2020**, *395*, 200–211. [CrossRef] [PubMed]
3. Machado, F.R.; Cavalcanti, A.B.; Bozza, F.A.; Ferreira, E.M.; Sousa, F.; Carrara, A.; Sousa, J.L. The Epidemiology of Sepsis in Brazilian Intensive Care Units (the Sepsis PREvalence Assessment Database, SPREAD): An Observational Study. *Lancet Infect. Dis.* **2017**, *17*, 1180–1189. [CrossRef] [PubMed]
4. Evans, L.; Rhodes, A.; Alhazzani, W.; Antonelli, M.; Coopersmith, C.M.; French, C.; Machado, F.R.; Mcintyre, L.; Ostermann, M.; Prescott, H.C.; et al. Surviving Sepsis Campaign: International Guidelines for Management of Sepsis and Septic Shock 2021. *Intensive Care Med.* **2021**, *47*, 1181–1247. [CrossRef] [PubMed]
5. Nguyen, H.B.; Jaehne, A.K.; Jayaprakash, N.; Semler, M.W.; Hegab, S.; Yataco, A.C.; Tatem, G.; Salem, D.; Moore, S.; Boka, K.; et al. Early Goal-Directed Therapy in Severe Sepsis and Septic Shock: Insights and Comparisons to ProCESS, ProMISe, and ARISE. *Crit. Care* **2016**, *20*, 160. [CrossRef] [PubMed]
6. Lambden, S.; Laterre, P.F.; Levy, M.M.; Francois, B. The SOFA Score—Development, Utility and Challenges of Accurate Assessment in Clinical Trials. *Crit Care* **2019**, *23*, 374. [CrossRef] [PubMed]
7. Shankar-Hari, M.; Phillips, G.S.; Levy, M.L.; Seymour, C.W.; Liu, V.X.; Deutschman, C.S.; Angus, D.C.; Rubenfeld, G.D.; Singer, M. Developing a Newdefinition and Assessing Newclinical Criteria for Septic Shock: For the Third International Consensus Definitions for Sepsis and Septic Shock (Sepsis-3). *JAMA J. Am. Med. Assoc.* **2016**, *315*, 775–787. [CrossRef] [PubMed]
8. Wang, C.; Xu, R.; Zeng, Y.; Zhao, Y.; Hu, X. A Comparison of QSOFA, SIRS and NEWS in Predicting the Accuracy of Mortality in Patients with Suspected Sepsis: A Meta- Analysis. *PLoS ONE* **2022**, *82*, e0266755. [CrossRef] [PubMed]
9. Adegbite, B.R.; Edoa, J.R.; Ndoumba, W.F.N.; Mbadinga, L.B.D.; Mombo-Ngoma, G.; Jacob, S.T.; Rylance, J.; Hänscheid, T.; Adegnika, A.A.; Grobusch, M.P. A Comparison of Different Scores for Diagnosis and Mortality Prediction of Adults with Sepsis in Low-and-Middle-Income Countries: A Systematic Review and Meta-Analysis. *EClinicalMedicine* **2021**, *42*, 101184. [CrossRef]
10. Charoensappakit, A.; Sae-khow, K.; Rattanaliam, P.; Vutthikraivit, N.; Pecheenbuvan, M.; Udomkarnjananun, S.; Leelahavanichkul, A. Cell-Free DNA as Diagnostic and Prognostic Biomarkers for Adult Sepsis: A Systematic Review and Meta-Analysis. *Sci. Rep.* **2023**, *13*, 19624. [CrossRef]
11. de Miranda, F.S.; Barauna, V.G.; Dos Santos, L.; Costa, G.; Vassallo, P.F.; Campos, L.C.G. Properties and Application of Cell-Free DNA as a Clinical Biomarker. *Int. J. Mol. Sci.* **2021**, *22*, 9110. [CrossRef] [PubMed]
12. Brazil. RESOLUTION No. 466, of 12 DECEMBER 2012. 2012, No. 466. Available online: https://conselho.saude.gov.br/resolucoes/2012/466_english.pdf (accessed on 5 March 2023).
13. UN. Declaration of Helsinki. Recommendations Guiding Doctors in Clinical Research. *WHO Chron.* **1965**, *19*, 31–32. [CrossRef]
14. Tug, S.; Helmig, S.; Deichmann, E.R.; Schmeier-Jürchott, A.; Wagner, E.; Zimmermann, T.; Radsak, M.; Giacca, M.; Simon, P. Exercise-Induced Increases in Cell Free DNA in Human Plasma Originate Predominantly from Cells of the Haematopoietic Lineage. *Exerc. Immunol. Rev.* **2015**, *21*, 164–173.
15. Tug, S.; Tross, A.; Hegen, P.; Wanja, E.; Neuberger, I.; Simon, P.; Helmig, S.; Scho, W. Acute Effects of Strength Exercises and Effects of Regular Strength Training on Cell Free DNA Concentrations in Blood Plasma. *PLoS ONE* **2017**, *12*, e0184668. [CrossRef]
16. Beiter, T.; Fragasso, A.; Hudemann, J.; Nieß, A.M.; Simon, P. Short-Term Treadmill Running as a Model for Studying Cell-Free DNA Kinetics in Vivo. *Clin. Chem.* **2011**, *57*, 633–636. [CrossRef] [PubMed]
17. Meddeb, R.; Dache, Z.A.A.; Thezenas, S.; Otandault, A.; Tanos, R.; Pastor, B.; Sanchez, C.; Azzi, J.; Tousch, G.; Azan, S.; et al. Quantifying Circulating Cell-Free DNA in Humans. *Sci. Rep.* **2019**, *9*, 5220. [CrossRef] [PubMed]
18. Barichello, T.; Generoso, J.S.; Singer, M.; Dal-Pizzol, F. Biomarkers for Sepsis: More than Just Fever and Leukocytosis—A Narrative Review. *Crit. Care* **2022**, *26*, 14. [CrossRef]
19. Pierrakos, C.; Velissaris, D.; Bisdorff, M.; Marshall, J.C.; Vincent, J. Biomarkers of Sepsis: Time for a Reappraisal. *Crit. Care* **2020**, *24*, 287. [CrossRef]
20. Pierrakos, C.; Vincent, J.L. Sepsis Biomarkers: A Review. *Crit. Care* **2010**, *14*, R15. [CrossRef]
21. Bronkhorst, A.J.; Ungerer, V.; Diehl, F.; Anker, P.; Dor, Y.; Fleischhacker, M.; Gahan, P.B.; Hui, L.; Holdenrieder, S.; Thierry, A.R. Towards Systematic Nomenclature for Cell-Free DNA. *Hum. Genet.* **2021**, *140*, 565–578. [CrossRef]
22. Zeng, X.; Li, X.; Zhang, Y.; Cao, C.; Zhou, Q. IL6 Induces MtDNA Leakage to Affect the Immune Escape of Endometrial Carcinoma via CGAS-STING. *J. Immunol. Res.* **2022**, *2022*, 3815853. [CrossRef]
23. Harrington, J.S.; Choi, A.M.K.; Nakahira, K. Mitochondrial DNA in Sepsis. *Curr. Opin. Crit. Care* **2017**, *23*, 284–290. [CrossRef] [PubMed]
24. Clementi, A.; Virzì, G.M.; Brocca, A.; Pastori, S.; De Cal, M.; Marcante, S.; Granata, A.; Ronco, C. The Role of Cell-Free Plasma DNA in Critically Ill Patients with Sepsis. *Blood Purif.* **2016**, *41*, 34–40. [CrossRef] [PubMed]
25. Duplessis, C.; Gregory, M.; Frey, K.; Bell, M.; Truong, L.; Schully, K.; Lawler, J.; Langley, R.J.; Kingsmore, S.F.; Woods, C.W.; et al. Evaluating the Discriminating Capacity of Cell Death (Apoptotic) Biomarkers in Sepsis. *J. Intensive Care* **2018**, *5*, 72. [CrossRef] [PubMed]

26. Rannikko, J.; Seiskari, T.; Huttunen, R.; Tarkiainen, I.; Jylhävä, J.; Hurme, M.; Syrjänen, J.; Aittoniemi, J. Plasma Cell-Free DNA and QSOFA Score Predict 7-Day Mortality in 481 Emergency Department Bacteraemia Patients. *J. Intern. Med.* **2018**, *284*, 418–426. [CrossRef]
27. Timmermans, K.; Kox, M.; Scheffer, G.J.; Pickkers, P. Plasma Nuclear and Mitochondrial Dna Levels, and Markers of Inflammation, Shock, and Organ Damage in Patients with Septic Shock. *Shock* **2016**, *45*, 607–612. [CrossRef]
28. Schäfer, S.T.; Adamzik, M.; Schumak, B.; Scherag, A.; Engler, A.; Schönborn, N.; Walden, J.; Koch, S.; Baba, H.A.; Steinmann, J. Mitochondrial DNA. *Anesthesiology* **2016**, *124*, 923–933. [CrossRef]
29. Bhagirath, V.C.; Dwivedi, D.J.; Liaw, P.C. Comparison of the Proinflammatory and Procoagulant Properties of Nuclear, Mitochondrial, and Bacterial DNA. *Shock* **2015**, *44*, 265–271. [CrossRef]
30. Devonshire, A.S.; Whale, A.S.; Gutteridge, A.; Jones, G.; Cowen, S.; Foy, C.A.; Huggett, J.F. Towards Standardisation of Cell-Free DNA Measurement in Plasma: Controls for Extraction Efficiency, Fragment Size Bias and Quantification. *Anal. Bioanal. Chem.* **2014**, *406*, 6499–6512. [CrossRef]
31. Moss, J.; Magenheim, J.; Neiman, D.; Zemmour, H.; Loyfer, N.; Korach, A.; Samet, Y.; Maoz, M.; Druid, H.; Arner, P.; et al. Comprehensive Human Cell-Type Methylation Atlas Reveals Origins of Circulating Cell-Free DNA in Health and Disease. *Nat. Commun.* **2018**, *9*, 5068. [CrossRef]
32. Rabaan, A.A.; Bakhrebah, M.A.; Alotaibi, J.; Natto, Z.S.; Alkhaibari, R.S.; Alawad, E.; Alshammari, H.M.; Alwarthan, S.; Alhajri, M.; Almogbel, M.S.; et al. Unleashing the Power of Artificial Intelligence for Diagnosing and Treating Infectious Diseases: A Comprehensive Review. *J. Infect. Public Health* **2023**, *16*, 1837–1847. [CrossRef] [PubMed]
33. O'Reilly, D.; McGrath, J.; Martin-Loeches, I. Optimizing Artificial Intelligence in Sepsis Management: Opportunities in the Present and Looking Closely to the Future. *J. Intensive Med.* **2023**, *4*, 34–45. [CrossRef] [PubMed]

biomedicines

MDPI

Article

Analysis of Calprotectin as an Early Marker of Infections Is Economically Advantageous in Intensive Care-Treated Patients

Aleksandra Havelka [1,2], Anders O. Larsson [3], Johan Mårtensson [4,5], Max Bell [4,5], Michael Hultström [6,7,8,9], Miklós Lipcsey [6,10] and Mats Eriksson [6,11,*]

1 Department of Molecular Medicine and Surgery, Karolinska Institute, 171 76 Stockholm, Sweden; aleksandra.havelka@gentian.no
2 Gentian Diagnostics AS, 1596 Moss, Norway
3 Department of Medical Sciences, Clinical Chemistry, Uppsala University, 751 85 Uppsala, Sweden
4 Department of Physiology and Pharmacology, Section of Anaesthesiology and Intensive Care Medicine, Karolinska Institute, 171 77 Stockholm, Sweden
5 Department of Perioperative Medicine and Intensive Care, Karolinska University Hospital, 171 76 Stockholm, Sweden
6 Department of Surgical Sciences, Section of Anaesthesiology and Intensive Care Medicine, Uppsala University, 751 85 Uppsala, Sweden
7 Department of Medical Cell Biology, Integrative Physiology, Uppsala University, 751 23 Uppsala, Sweden
8 Department of Epidemiology, McGill University, Montréal, QC H3A 0G4, Canada
9 Lady Davis Institute of Medical Research, Jewish General Hospital, Montréal, QC H3T 1E2, Canada
10 Hedenstierna Laboratory, Department of Surgical Sciences, Uppsala University, 751 85 Uppsala, Sweden
11 NOVA Medical School, New University of Lisbon, 1099-085 Lisbon, Portugal
* Correspondence: mats.eriksson@surgsci.uu.se; Tel.: +46-18-6110000 or +46-76-1-333-818

Abstract: Calprotectin is released from neutrophil granulocytes upon activation. Several studies have indicated that plasma calprotectin is an early determinant of bacterial infections, which may serve as a diagnostic tool facilitating decision making on antibiotic treatment. The study objective was to explore the health and economic implications of calprotectin as a predictive tool to initiate antimicrobial therapy in a cohort of critically ill patients. Thus, data obtained from a previously published study on calprotectin as a hypothetical early biomarker of bacterial infections in critically ill patients were evaluated regarding the potential cost-effective impact of early analysis of calprotectin on an earlier start of antibiotic treatment. Under the assumption that calprotectin is used predictively and comparators (white blood cells, procalcitonin, and C-reactive protein) are used diagnostically, a cost-effective impact of EUR 11,000–12,000 per patient would be obtained. If calprotectin would be used predictively and comparators would be used predictively for 50% of patients, it is hypothesized that cost-effectiveness would be between EUR 6000 and 7000 per patient, based on reduced stay in the ICU and general ward, respectively. Furthermore, predictive use of calprotectin seems to reduce both mortality and the length of hospital stay. This health economic analysis on the predictive use of plasma calprotectin, which facilitates clinical decision making in cases of suspected sepsis, indicates that such determination has a cost-saving and life-saving impact on the healthcare system.

Keywords: calprotectin; costs; early detection; economic modeling; infection; intensive care; sepsis

Citation: Havelka, A.; Larsson, A.O.; Mårtensson, J.; Bell, M.; Hultström, M.; Lipcsey, M.; Eriksson, M. Analysis of Calprotectin as an Early Marker of Infections Is Economically Advantageous in Intensive Care-Treated Patients. *Biomedicines* **2023**, *11*, 2156. https://doi.org/10.3390/biomedicines11082156

Academic Editors: Wen-Lin Su, Sheng-Kang Chiu, Chih-Hao Shen and Yi-Ting Chen

Received: 3 July 2023
Revised: 28 July 2023
Accepted: 30 July 2023
Published: 1 August 2023

1. Introduction

Globally, the estimated number of sepsis cases in 2017 was nearly 50 million, and there were approximately 11 million sepsis-related deaths [1]. Deaths in severe sepsis and septic shock are not restricted to the acute phase but remain high even three months after diagnosis [2]. One of the most important aspects of the management of patients with sepsis is early recognition so that the administration of antibiotics, source control measures, and effective resuscitation strategies can be initiated as soon as possible [3].

Prompt administration of antibiotics is associated with improved outcomes in sepsis and septic shock. Every 1 h delay in antibiotics after emergency department (ED) triage or the onset of organ dysfunction or shock may lead to a 3–7% increase in the odds of a poor outcome [4–6]. A quality improvement program has resulted in a significant reduction in in-hospital sepsis cases and sepsis mortality and yielded a considerable positive return on investment [7], showing that improvements can be achieved. The mean total hospital costs per patient with sepsis vary considerably between countries. The relative amount of healthcare budget spent on sepsis per country is estimated at 2.65%, which equals a median (interquartile range (IQR)) amount of gross national product spent on sepsis-related healthcare of 0.33% (0.03–2.27%) [8]. Early identification of patients with sepsis is a prerequisite to triggering healthcare interventions to improve outcomes. Cost-effective biomarkers with rapidly available results around the clock and high specificity and sensitivity could improve the management of septic patients.

In severe infections, neutrophil granulocytes adhere to and penetrate vascular walls, which reduces their number in the blood and reduces the predictive value of the number of circulating neutrophil granulocytes [9]. Calprotectin, a heterodimer of two protein subunits, is present in high concentrations in the cytoplasm of neutrophils and is released into the circulation upon neutrophil activation without the requirement for de novo synthesis.

Several studies have confirmed the fast release of calprotectin upon inflammatory stimuli and earlier detection of calprotectin in circulation compared to procalcitonin (PCT) and C-reactive protein (CRP) [10,11].

Calprotectin release weakens cellular adhesion, favoring leukocyte extravasation, and activates pro-inflammatory pathways regulated by the cell surface receptors TLR 4 and RAGE, leading to the expression of pro-inflammatory cytokines and amplification of the inflammatory response. CRP expression is induced by interleukin-6 (IL-6) and occurs in the liver, which results in slower kinetics and late detection of this biomarker in the circulation. IL-6 is rapidly but transiently expressed in response to infection, inflammation, and tissue injury [12]. Calprotectin has been shown to be a valuable biomarker for early detection of infections [9–11], and its soluble form provides both bacteriostatic and cytokine-like effects in the local environment [13].

Calprotectin, analyzed at admission to the intensive care unit (ICU), has a high sensitivity for bacterial infections in severely ill patients and has predictive value for mortality [14–16]. Previous studies confirm the value of calprotectin in the identification of bacterial infections, with better sensitivity and specificity than CRP and PCT [14,15,17,18]. Moreover, calprotectin was superior to PCT when patients with and without bacterial sepsis were distinguished [15].

Based on the results from previous studies, we hypothesized that identifying patients with severe infections/sepsis at an early stage using calprotectin would be beneficial for patients and healthcare providers and economically superior to other commonly used biomarkers.

Consequently, the primary aim of the present study was to evaluate whether analysis of calprotectin might be an economically beneficial complement to conventionally used biomarkers of infection, i.e., CRP, white blood cell count (WBC), and PCT, in a previously published cohort of unselected ICU patients [16], where the aim was to assess plasma calprotectin as an early marker of bacterial infections in critically ill patients in comparison with PCT, CRP, and WBC.

2. Materials and Methods

2.1. Clinical Setting and Target Population

This health economic analysis is based on a study published by Jonsson et al. [16], aiming to assess the value of plasma calprotectin as an early biomarker of bacterial infections in critically ill patients. One hundred and eighty-eight eligible patients with an expected length of stay (LOS) > 24 h admitted to the General Intensive Care Unit (ICU) at Karolinska University Hospital, Solna, Sweden, were evaluated for inclusion in the study. A flowchart showing the patient selection path has previously been published [16], showing 110 included patients, out of which 58 patients (52.7%) had antibiotic therapy initiated during their ICU stay. To avoid potential misclassification in our analysis, we excluded 20 patients with a possible or probable infection. Therefore, we included 58 patients with no infection and 38 patients with a confirmed infection in the final analysis. Initial blood samples were collected on ICU admission. The results of the calprotectin analyses were blinded to an infectious disease specialist who determined the likelihood of infection.

In the present hypothetical study on the cost-effectiveness of calprotectin, calculations were based on the findings of Jonsson et al. [16]. Their study was approved by The Regional Ethics Review Board in Stockholm, Sweden (2006/1469-31/2 and subsequent amendments) and performed in accordance with ethical principles that have their origin in the Declaration of Helsinki [19], which are consistent with ICH/Good Clinical Practice (GCP) E6 (R2), and EU Clinical Trials Directive, and applicable local regulatory requirements. Since the present study consists of de-identified data in which patients could not be individually identified during the analysis process; patient consent was not required. Our results are based on a theoretical model using an algorithm for mathematical evaluation, without any new collection of clinical data. Hence, renewed ethical permission was not necessary [20].

2.2. Analyses of Biomarkers

At ICU admission, or shortly thereafter, and on consecutive days, up to one week, blood samples were collected in EDTA tubes and centrifuged at 2000 rpm for 10 min at +4 °C. Supernatants were stored at −80 °C until analyzed at the Department of Clinical Chemistry, Uppsala University Hospital, Uppsala, Sweden. Calprotectin was analyzed using a Mindray BS-380 (Mindray Medical International, Shenzhen, China) and a particle-enhanced turbidimetric immunoassay, (GCAL® Calprotectin Immunoassay) from Gentian AS (Moss, Norway). CRP was analyzed on an Architect Ci8200 (Abbott Laboratories, Abbott Park, IL, USA). The expected normal CRP level was <5 mg/L. PCT was analyzed using Cobas EE (Roche Diagnostics, Mannheim, Germany). The expected normal PCT level was <0.05 ng/mL.

2.3. Base Data and Model Structure

A decision tree model is intended to capture the initial outcomes of two or more decisions without considering time as a component of the outcome of the decisions. Since the intervention (calprotectin analysis) and the comparators have the same intended outcomes within a limited period of time, a decision tree model was chosen as the most appropriate for calculating the cost-effectiveness of calprotectin.

The decision tree shows the different diagnostic outcomes of correctly or incorrectly diagnosing a bacterial infection. The same cohort of patients is subjected to either detecting or not detecting the infection, which determines the timing of the antibiotic treatment and the outcomes (survival, length of hospital stay, and costs). The sensitivity of the tests determines the proportion of patients who are correctly (true positive) and incorrectly (false positive) tested for having a bacterial infection, and the specificity determines those who are correctly and incorrectly diagnosed as not having a bacterial infection (true negative and false negative, respectively). The structure of the decision tree is presented in Figure 1.

The patient pathway in the decision tree for a predictive test starts with defining the population (patients admitted to the ICU); however, this does not in itself impact the pathway. The sensitivity and specificity of the predictive biomarker test will determine what proportion of the patient population is correctly or incorrectly diagnosed. The proportion of those who are correctly diagnosed will be predictively detected with a bacterial infection and receive antibiotic treatment 24 h prior to clinical presentation and the subsequent antibiotic treatment. The proportion correctly diagnosed as not having a bacterial infection (true negative) will not be treated with antibiotics. The proportion of the patient population that is incorrectly diagnosed will not be detected before clinical presentation and, thus, will receive antibiotic treatment after clinical presentation. The patient population that is incorrectly diagnosed where the predictive test suggests bacterial infection (false positive) will receive antibiotic treatment together with the patients who are predictively detected correctly. Based on the pathway, patients face varying outcomes (mortality, length of stay), resulting in different costs and survival rates.

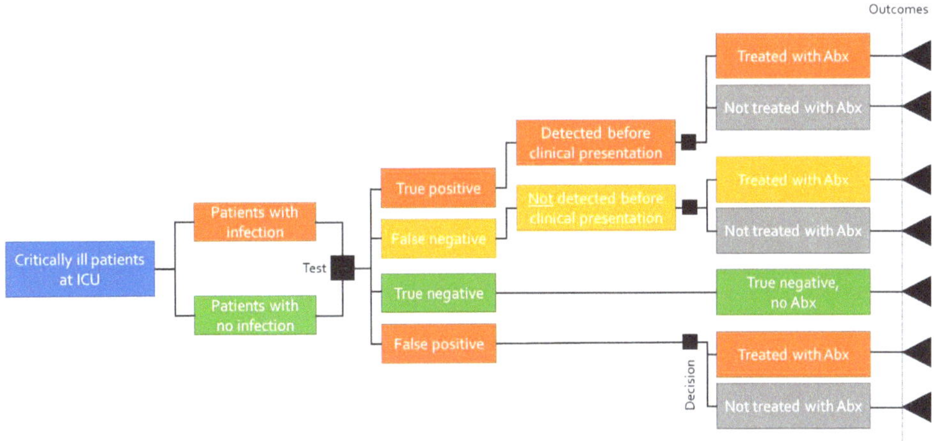

Figure 1. Decision tree structure with a predictive test.

Figure 2 illustrates the decision tree structure when there is no test taken. In this case, the proportion of patients with an infection will not be detected before clinical presentation, which will result in a longer length of stay and an increased risk of mortality.

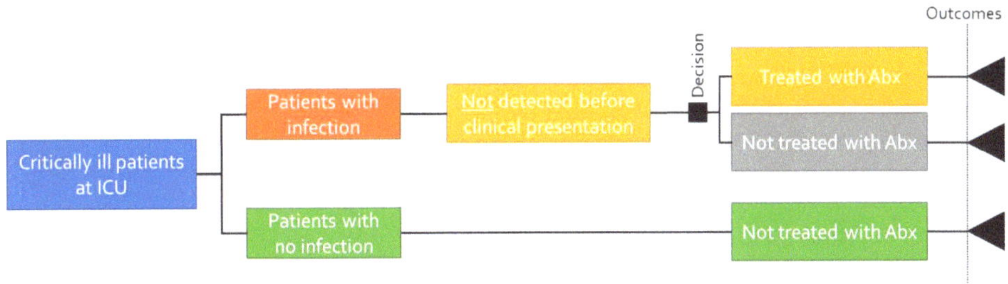

Figure 2. Decision tree structure with no diagnostic test.

This model will estimate the cost-effectiveness of calprotectin analysis for early detection of bacterial infections as compared to analysis of WBC, procalcitonin, CRP, and/or conducting no test at all in a Swedish intensive care setting. Outcomes include costs, mortality rates, and length of stay at hospital. The timing of the first administration of antibiotics affects the total length of stay in the hospital [6]. In this scenario, it is assumed that early detection of sepsis implies a total hospital stay of 15 days, while symptomatic detection is associated with 30 days of hospitalization.

The comparators included in the analysis are WBC, PCT, CRP, and no testing. The analysis is based on patients admitted to an ICU in Sweden, with inclusion and exclusion criteria from the literature applied [16].

Table 1 presents the predictive accuracy of detecting a bacterial infection by the respective biomarker 24 h before initiating antibiotic treatment. The diagnostic accuracy of the biomarker test is presented in Table 2. The diagnostic test is taken on the same day as the initiation of the antibiotic treatment when the patient is symptomatic.

Table 1. Sensitivity and specificity of predictive test (24 h prior to antibiotic prescription).

Biomarker (Test)	Sensitivity	Specificity	Source
Calprotectin immunoassay	66%	93%	[16]
WBC count	67%	41%	
Procalcitonin	58%	58%	
C-reactive protein (CRP)	58%	94%	
No test	0%	0%	Assumption

Table 2. Sensitivity and specificity of diagnostic test (same day as antibiotic prescription).

Biomarker (Test)	Sensitivity	Specificity	Source
Calprotectin immunoassay	55.0%	90.0%	[16]
White blood cell count	42.0%	74.0%	
Procalcitonin	73.0%	58.0%	
C-reactive protein (CRP)	82.0%	54.0%	
No test	0.0%	0.0%	Assumption

As evaluated by the areas under the receiver-operating characteristics (ROC) curves, calprotectin predicted bacterial infections at 0.82 (95% CI, 0.70–0.94), whereas all the other biomarkers included in the study had lower ROC areas [16].

2.4. Diagnostic Yield

The diagnostic yield is the probability of a patient having a positive diagnosis, i.e., a patient is diagnosed to have a bacterial infection and needs antibiotic treatment. This does not only depend on the sensitivity and specificity outlined above but also on the number of sequential tests performed, whether the test is taken predictively or diagnostically, and if the physician decides to initiate antibiotic treatment or not after analyzing the test results.

The model allows for a total of 10 sequential tests, distributed as either predictive or diagnostic. Since the predictive performance of GCAL® is central to the CE analysis, 100% of patients in the GCAL arm will undergo predictive tests and 0% in the comparator arms (which can be changed between 0 and 100% by the user). The test sensitivity and specificity are adjusted accordingly to the proportion of patients who are assumed to use the test predictively, i.e., in the base case, no patients are identified predictively for the comparators. The estimated treatment cost per day in the ICU is EUR 5530 and EUR 591 in a general ward (see "Limitations" section).

The model will by default assume that the physician will trust the test if the test result suggests a bacterial infection and initiate the antibiotic treatment early. In cases where tests are used predictively and the test result is negative, another test will be taken when the patient becomes symptomatic of infection. The model assumes by default that the physician will perform a diagnostic test and initiate antibiotic treatment if the patient is symptomatic. This confirmatory test affects the overall sensitivity and specificity.

The base-case settings and the diagnostic yield result (final sensitivity and specificity) for the base case are presented in Table 3. Notice the comparators remain equal to Table 2 as they are used diagnostically; however, the calprotectin values are updated due to the possibility of receiving a false negative result (as previously described).

Table 3. Applied sensitivity and specificity in the base-case scenario.

Biomarker (Test)	Sensitivity	Specificity	Source
Calprotectin immunoassay	85%	99%	Calculations
WBC count	42%	74%	
Procalcitonin	73%	58%	
C-reactive protein (CRP)	82%	54%	
No test	0.0%	0.0%	Assumption

The length of the ICU stay was estimated by an algorithm defined by Ferrer et al. [6] and Paoli et al. [21], in which they stratified the unadjusted length of stay (LoS) in an ICU by the timing of the antibiotic therapy. In patients presenting with sepsis in the emergency department, the linear relationship is as follows: LoS = 8.4143 + 0.7714 × hours to antibiotic therapy, whereas mortality (%) was equal to 0.2291 + 0.0241 × hours to antibiotic therapy [21]. However, the Ferrer et al. study [6] only includes data on antibiotic administration from 0 to 6+ h after admission to an ICU, and no hour-specific data were presented after 6 h. It is also important to note that the Ferrer et al. data [6] suggest that patients who received treatment in the first hour are most likely to have more severe symptoms present. Therefore, the LoS algorithm has a lower bound of 1 h and an upper bound of 6 h for antibiotic treatment. As the Ferrer et al. study [6] was not performed in a Swedish setting, an adjustment of the equation estimating LoS was considered necessary. Thus, we applied a multiplier of 0.5 in order to utilize a safety margin.

Symptomatic detection is assumed to be equal to the upper bound of the Ferrer algorithm [6], and thus, early detection is assumed to be 6 h before symptomatic detection. The LoS for early detection of infection and symptomatic detection following the algorithm and applied adjustment are presented in Table 4.

Table 4. Length of stay at ICU with the mathematical linear relationship.

	Default Days	Source
Early detection (6 h earlier)	4.2	[6,21]
Symptomatic detection	6.5	
	Share	Source
Adjustment of LoS	0.50	Assumed safety margin

3. Results

In the base-case scenario, calprotectin is employed predictively, differently from the comparators that are employed diagnostically. Under these assumptions, it is suggested that the analysis of calprotectin in patients with suspected sepsis saves costs and reduces in-hospital mortality in those patients. The mean duration of in-patient care is consequently reduced compared to all comparators (Table 5a,b).

Currently, no alternative settings for various inputs have been identified besides the severity adjustments available in the model. Thus, arbitrary ranges of assumed key inputs have been used to assess the impact these inputs have on the model results.

When including a predictive test for 50% of patients in the comparator scenarios (except the no-test option), analysis of calprotectin remains the dominant option (Table 6a,b).

Table 5. (a) Base-case outcomes. (b) Base-case outcomes (difference compared to calprotectin) per comparator.

(a)				
	Total Cost [EUR (Rounded off)]	**Deaths**	**Mean Days ICU**	**Mean Days General Ward**
Calprotectin immunoassay	25,000	0.2	3.7	8.0
WBC count	36,000	0.3	5.0	15.1
Procalcitonin	37,000	0.3	5.1	15.3
C-reactive protein (CRP)	37,000	0.3	5.2	14.9
No test	40,000	0.3	5.7	15.8
(b)				
	Δ Total Cost [EUR (Rounded off)]	**Δ Deaths**	**Δ Mean days ICU**	**Δ Mean Days Ward**
Calprotectin immunoassay	Reference	Reference	Reference	Reference
WBC count	11,000	0.1	+1.3	+7.1
Procalcitonin	12,000	0.1	+1.4	+7.3
C-reactive protein (CRP)	12,000	0.1	+1.5	+6.9
No test	15,000	0.1	+2.0	+7.8

Table 6. (a) Outcomes when using comparators predictively for 50% of patients instead of 0%. (b) Outcomes (difference compared to calprotectin) per comparator when using comparators predictively for 50% of patients instead of 0%.

(a)				
	Total Cost [EUR (Rounded off)]	**Deaths**	**Mean Days ICU**	**Mean Days General Ward**
Calprotectin immunoassay	25,000	0.2	3.7	8.0
WBC count	32,000	0.3	4.5	13.0
Procalcitonin	32,000	0.3	4.5	12.5
C-reactive protein	31,000	0.3	4.4	11.9
No test	40,000	0.3	5.7	15.8
(b)				
	Δ Total Cost [EUR (Rounded off)]	**Δ Deaths**	**Δ Mean Days ICU**	**Δ Mean Days Ward**
Calprotectin immunoassay	Reference	Reference	Reference	Reference
WBC count	7000	+0.1	+0.8	+5.0
Procalcitonin	7000	+0.1	+0.8	+4.5
C-reactive protein	6000	−0.1	+0.7	+3.9
No test	15,000	+0.1	+1.9	+7.8

The calprotectin assay has been shown to detect infection up to 24 h prior to symptomatic onset [16]. However, the linear relationship in the Ferrer et al. algorithm [6] cannot be extended beyond the 6 h difference, as doing so equates to a negative LoS for 24 h early detection. Therefore, to assess the impact of earlier mean time to antibiotics by determining plasma calprotectin, the LoS at an ICU for early detection is varied from 0 days to a maximum of 6.5 days, which is equal to the LoS for symptomatic detection. Table 7 displays the results of these analyses. No changes in the comparators are observed, as they are not used predictively in these cases. Thus, they were not included in the table.

Table 7. Variation in LoS at an ICU for early detection using GCAL®.

Early Detection LoS in an ICU	Total Cost [EUR (Rounded off)]	Deaths	Mean Days ICU	Mean Days General Ward
0 days	15,000	0.2	1.8	9.9
1 day	17,000	0.2	2.3	9.4
2 days	20,000	0.2	2.7	9.0
3 days	22,000	0.2	3.2	8.5
4 days	24,000	0.2	3.6	8.1
5 days	26,000	0.2	4.0	7.7
6 days	28,000	0.2	4.5	7.2
6.5 days [equal to symptomatic (Table 4)]	29,000	0.2	4.7	7.0

Estimated differences in LoS in a general ward (G. ward) for early and symptomatic detection are shown in Tables 8 and 9, respectively. In these assumptions, the total length of in-patient care for early and symptomatic patients is 15 days and 30 days, respectively. These total LoS equate to 10.8 days and 23.5 days spent in a general ward for early and symptomatic treated patients. As these values are inherently uncertain, sensitivity analysis surrounding the total LoS and the related LoS in a general ward is assessed. Note that the LoS in an ICU remains the same as originally assumed. Analysis of calprotectin remains the principal option.

Table 8. Total LoS of 15 days for early and symptomatically treated patients.

	Total Cost [EUR (Rounded off)]	Deaths	Mean Days ICU	Mean Days General Ward
Calprotectin immunoassay	24,000	0.2	3.7	6.8
WBC count	31,000	0.3	4.9	7.2
Procalcitonin	32,000	0.3	5.1	7.4
C-reactive protein	32,000	0.3	5.2	6.9
No test	35,000	0.3	5.7	7.9

Table 9. Total LoS of 30 days for early and symptomatically treated patients.

	Total Cost [EUR (Rounded off)]	Deaths	Mean Days ICU	Mean Days General Ward
Calprotectin immunoassay	29,000	0.2	3.7	14.7
WBC count	36,000	0.3	4.9	15.1
Procalcitonin	37,000	0.3	5.1	15.3
C-reactive protein	37,000	0.3	5.2	14.9
No test	40,000	0.3	5.7	15.8

4. Discussion

This analysis, which was performed by an independent company [22], specializing in health economic research, is essentially based on a study indicating that plasma calprotectin is useful as an early complementary biomarker of bacterial infections in ICU patients [16], where WBC, PCT, and CRP may still provide valuable information. Hence, the advantage of calprotectin in this context assumes a fast increase in this biomarker in the circulation and, thereby, earlier detection of infection when compared to other routinely used biomarkers. Several other studies have also concluded that early detection of bacterial infections facilitates decision making on antimicrobial therapeutic interventions, which may improve outcomes [4,5,11,14,15,17,18,23,24]. Although this study focuses on a health economic perspective, the main rationale for the analysis of calprotectin is from a clinical perspective, since early diagnosis of severe infections and sepsis reduces both delays in treatment and mortality [5,24–26]. From this aspect, our findings are in alignment with previous ones, indicating that early detection of severe infections and sepsis has both a cost-saving and a life-saving impact in the ICU setting [7,21].

An estimated median (IQR) total hospital cost per patient stay for sepsis is EUR 32,421 (USD 20,090), depending on the country and type of hospital. The median reported hospital-wide costs (IQR) of survivors and non-survivors were EUR 34,855 (USD 14,352) and EUR 20,537 (USD 17,817), respectively [25] (USD/EUR ~0.95). In a recent study on cost-effectiveness in the ICU, it was deduced that if sepsis could be detected 3 h earlier than current practice, this would result in substantial annual cost savings for the healthcare system [26].

In a U.S. cohort of patients developing sepsis during hospital stay but not diagnosed on admission, the mean overall cost was more than EUR 51,022, whereas the mean cost for those diagnosed with sepsis on admission was EUR 18,023 [23]. When sepsis was not diagnosed on admission, the mortality rate was twice that seen when sepsis was diagnosed on admission. However, it cannot be excluded that higher costs and a poorer prognosis, at least to some extent, may be attributable to some patients developing sepsis during the hospital stay.

Calprotectin has also been shown to be a promising biomarker for estimation of disease severity, prediction of organ failure, and mortality in patients with severe infections, including bacterial infections and SARS-CoV-2 infections presented at the emergency department as well as in an ICU setting [24,27–29]. Unlike other routinely used inflammatory biomarkers, such as CRP and PCT, calprotectin is released by neutrophils into the bloodstream without requiring de novo protein biosynthesis [27]. Calprotectin can today be measured on high-throughput instruments at the same time to result in CRP, which is important for acute diagnoses and optimal treatment decisions.

Strengths and Limitations

Although this is a retrospective study, the original data [16] were collected prospectively. Calprotectin was sampled daily and was significantly elevated in infected patients. Since therapeutic interventions were not steered by calprotectin levels, this indicates that the prediction of calprotectin as a biomarker of bacterial infection was unbiased. We excluded patients with uncertain infection status, i.e., patients with a possible or probable infection, to avoid misclassification. However, such patients are common in the ICU and may also benefit from early antibiotic therapy.

The most obvious limitation of this study is that the results are due to the use of an algorithm. However, the used algorithm is based on international data sources [6,21,23] and with slight modifications to fit with Swedish clinical practice, which provides a more accurate estimate than simpler models. The assumption that calprotectin should have been taken predictively, whereas the comparators were analyzed based on clinical indications, is also a limitation. A better approach would have been a time-dependent analysis, which, however, is not possible retrospectively. Estimations of hospital costs were performed in a health economic analysis reported by Quantify Research AB, Stockholm, Sweden [22] and based on publicly available information ((in Swedish); SEK 1/EUR ~0.09). Since both hospital costs and exchange rates vary over time, calculated expenses are rounded off to the closest thousandfold number just below the estimated price.

It is a drawback that the diagnostic accuracy of calprotectin, measured in combination with conventionally analyzed biomarkers (i.e., WBC, PCT, and CRP), was not calculated. However, it is tempting to speculate that algorithms managed by artificial intelligence may be useful for such calculations in the not-too-distant future.

5. Conclusions

In the present base-case scenario, calprotectin was identified as a cost-effective biomarker for an unselected patient cohort presenting in a Swedish ICU. Compared to procalcitonin, leucocytes, and CRP, it is suggested that calprotectin saves total costs, reduces the mean duration of in-patient care, and reduces in-hospital mortality in those patients. In the sensitivity analysis, calprotectin remains the dominant option when key model inputs are varied.

Author Contributions: Conceptualization, A.H., A.O.L.; methodology, A.H., A.O.L., M.E.; writing—original draft preparation was mainly performed by A.H., M.E.; writing—review and editing, All authors participated in the revisions of the manuscript. All authors read, commented, and approved the final manuscript for publication. All authors have read and agreed to the published version of the manuscript.

Funding: This study was supported by the Uppsala University Hospital Research Fund.

Institutional Review Board Statement: The present study consists of de-identified data in which patients could not be individually identified during the analysis process; patient consent was not required. Our results are based on a theoretical model using an algorithm for mathematical evaluation, without any new collection of clinical data.

Informed Consent Statement: Patient consent was waived due to the use of an algorithm where de-identified data were used.

Data Availability Statement: This study is based on a previously published study [16] that was ethically approved.

Acknowledgments: Health economic research calculations, based on a scientific publication [16], were performed by Quantify Research AB, Stockholm, Sweden [22] and remunerated by Gentian Diagnostics AB, Stockholm.

Conflicts of Interest: Aleksandra Havelka is Chief Scientific Officer at Gentian Diagnostics AS and General Manager at Gentian Diagnostics AB (https://www.gentian.com/).

References

1. Rudd, K.E.; Johnson, S.C.; Agesa, K.M.; Shackelford, K.A.; Tsoi, D.; Kievlan, D.R.; Colombara, D.V.; Ikuta, K.S.; Kissoon, N.; Finfer, S.; et al. Global, regional, and national sepsis incidence and mortality, 1990–2017: Analysis for the Global Burden of Disease Study. *Lancet* **2020**, *395*, 200–211. [CrossRef] [PubMed]
2. Strandberg, G.; Walther, S.; Agvald Öhman, C.; Lipcsey, M. Mortality after Severe Sepsis and Septic Shock in Swedish Intensive Care Units 2008-2016-A nationwide observational study. *Acta Anaesthesiol. Scand.* **2020**, *64*, 967–975. [CrossRef]
3. Vincent, J.L.; Pereira, A.J.; Gleeson, J.; Backer, D. Early management of sepsis. *Clin. Exp. Emerg. Med.* **2014**, *1*, 3–7. [CrossRef]
4. Kumar, A.; Haery, C.; Paladugu, B.; Kumar, A.; Symeoneides, S.; Taiberg, L.; Osman, J.; Trenholme, G.; Opal, S.M.; Goldfarb, R.; et al. The duration of hypotension before the initiation of antibiotic treatment is a critical determinant of survival in a murine model of Escherichia coli septic shock: Association with serum lactate and inflammatory cytokine levels. *J. Infect. Dis.* **2006**, *193*, 251–258. [CrossRef]
5. Kumar, A.; Roberts, D.; Wood, K.E.; Light, B.; Parrillo, J.E.; Sharma, S.; Suppes, R.; Feinstein, D.; Zanotti, S.; Taiberg, L.; et al. Duration of hypotension before initiation of effective antimicrobial therapy is the critical determinant of survival in human septic shock. *Crit. Care Med.* **2006**, *34*, 1589–1596. [CrossRef]
6. Ferrer, R.; Martin-Loeches, I.; Phillips, G.; Osborn, T.M.; Townsend, S.; Dellinger, R.P.; Artigas, A.; Schorr, C.; Levy, M.M. Empiric antibiotic treatment reduces mortality in severe sepsis and septic shock from the first hour: Results from a guideline-based performance improvement program. *Crit. Care Med.* **2014**, *42*, 1749–1755. [CrossRef]
7. Khowaja, A.R.; Willms, A.J.; Krause, C.; Carriere, S.; Ridout, B.; Kennedy, C.; Young, E.; Mitton, C.; Kissoon, N.; Sweet, D.D. The Return on Investment of a Province-Wide Quality Improvement Initiative for Reducing In-Hospital Sepsis Rates and Mortality in British Columbia, Canada. *Crit. Care Med.* **2022**, *50*, e340–e350. [CrossRef]
8. van den Berg, M.; van Beuningen, F.E.; Ter Maaten, J.C.; Bouma, H.R. Hospital-related costs of sepsis around the world: A systematic review exploring the economic burden of sepsis. *J. Crit. Care* **2022**, *71*, 154096. [CrossRef]
9. Lerman, Y.V.; Kim, M. Neutrophil migration under normal and sepsis conditions. *Cardiovasc. Hematol. Disord. Drug Targets* **2015**, *15*, 19–28. [CrossRef]
10. Fullerton, J.N.; Havelka, A.; Segre, E.; De Maeyer, R.P.; Maini, A.A.; Gilroy, D.W. Kinetics of Calprotectin, Procalcitonin and C-Reactive Protein in Healthy Volunteers Administered Intravenous Endotoxin. Available online: https://ccforum.biomedcentral.com/articles/10.1186/s13054-020-2772-3 (accessed on 1 April 2023).

11. Lipcsey, M.; Hanslin, K.; Stålberg, J.; Smekal, D.; Larsson, A. The time course of calprotectin liberation from human neutrophil granulocytes after Escherichia coli and endotoxin challenge. *Innate Immun.* **2019**, *25*, 369–373. [CrossRef] [PubMed]
12. Tanaka, T.; Narazaki, M.; Kishimoto, T. IL-6 in inflammation, immunity, and disease. *Cold Spring Harb. Perspect. Biol.* **2014**, *6*, a016295. [CrossRef]
13. Stríz, I.; Trebichavský, I. Calprotectin—A pleiotropic molecule in acute and chronic inflammation. *Physiol Res.* **2004**, *53*, 245–253. [CrossRef]
14. Simm, M.; Söderberg, E.; Larsson, A.; Castegren, M.; Nilsen, T.; Eriksson, M.; Lipcsey, M. Performance of plasma calprotectin as a biomarker of early sepsis: A pilot study. *Biomark. Med.* **2016**, *10*, 811–818. [CrossRef] [PubMed]
15. Larsson, A.; Tydén, J.; Johansson, J.; Lipcsey, M.; Bergquist, M.; Kultima, K.; Mandic-Havelka, A. Calprotectin is superior to procalcitonin as a sepsis marker and predictor of 30-day mortality in intensive care patients. *Scand J. Clin. Lab. Investig.* **2020**, *80*, 156–161. [CrossRef]
16. Jonsson, N.; Nilsen, T.; Gille-Johnson, P.; Bell, M.; Martling, C.R.; Larsson, A.; Mårtensson, J. Calprotectin as an early biomarker of bacterial infections in critically ill patients: An exploratory cohort assessment. *Crit. Care Resusc.* **2017**, *19*, 205–213. [PubMed]
17. Bartáková, E.; Štefan, M.; Stráníková, A.; Pospíšilová, L.; Arientová, S.; Beran, O.; Blahutová, M.; Máca, J.; Holub, M. Calprotectin and calgranulin C serum levels in bacterial sepsis. *Diagn. Microbiol. Infect. Dis.* **2019**, *93*, 219–226. [CrossRef]
18. Havelka, A.; Sejersen, K.; Venge, P.; Pauksens, K.; Larsson, A. Calprotectin, a new biomarker for diagnosis of acute respiratory infections. *Sci. Rep.* **2020**, *10*, 4208. [CrossRef]
19. Wma Declaration of Helsinki—Ethical Principles for Medical Research Involving Human Subjects. Available online: https://www.wma.net/policies-post/wma-declaration-of-helsinki-ethical-principles-for-medical-research-involving-human-subjects/ (accessed on 23 March 2022).
20. Swedish Ethical Review Authority. Available online: https://etikprovningsmyndigheten.se/en/ (accessed on 30 March 2023).
21. Paoli, C.J.; Reynolds, M.A.; Coles, C.; Gitlin, M.; Crouser, E. Predicted Economic Benefits of a Novel Biomarker for Earlier Sepsis Identification and Treatment: A Counterfactual Analysis. *Crit. Care Explor.* **2019**, *1*, e0029. [CrossRef] [PubMed]
22. Quantify Research AB. Available online: https://quantifyresearch.com/ (accessed on 1 May 2023).
23. Paoli, C.J.; Reynolds, M.A.; Sinha, M.; Gitlin, M.; Crouser, E. Epidemiology and Costs of Sepsis in the United States-An Analysis Based on Timing of Diagnosis and Severity Level. *Crit. Care Med.* **2018**, *46*, 1889–1897. [CrossRef]
24. Parke, Å.; Unge, C.; Yu, D.; Sundén-Cullberg, J.; Strålin, K. Plasma calprotectin as an indicator of need of transfer to intensive care in patients with suspected sepsis at the emergency department. *BMC Emerg. Med.* **2023**, *23*, 16. [CrossRef]
25. Arefian, H.; Heublein, S.; Scherag, A.; Brunkhorst, F.M.; Younis, M.Z.; Moerer, O.; Fischer, D.; Hartmann, M. Hospital-related cost of sepsis: A systematic review. *J. Infect.* **2017**, *74*, 107–117. [CrossRef] [PubMed]
26. Ericson, O.; Hjelmgren, J.; Sjövall, F.; Söderberg, J.; Persson, I. The Potential Cost and Cost-Effectiveness Impact of Using a Machine Learning Algorithm for Early Detection of Sepsis in Intensive Care Units in Sweden. *J. Health Econ Outcomes Res.* **2022**, *9*, 101–110. [CrossRef] [PubMed]
27. Bauer, W.; Diehl-Wiesenecker, E.; Ulke, J.; Galtung, N.; Havelka, A.; Hegel, J.K.; Tauber, R.; Somasundaram, R.; Kappert, K. Outcome prediction by serum calprotectin in patients with COVID-19 in the emergency department. *J. Infect.* **2021**, *82*, 84–123. [CrossRef]
28. Cardiero, G.; Palma, D.; Vano, M.; Anastasio, C.; Pinchera, B.; Ferrandino, M.; Gianfico, C.; Gentile, L.; Savoia, M.; Gentile, I.; et al. Calprotectin Levels and Neutrophil Count Are Prognostic Markers of Mortality in COVID-19 Patients. *Diagnostics* **2022**, *12*, 2554. [CrossRef]
29. García de Guadiana-Romualdo, L.; Rodríguez Rojas, C.; Morell-García, D.; Andaluz-Ojeda, D.; Rodríguez Mulero, M.D.; Rodríguez-Borja, E.; Ballesteros-Vizoso, A.; Calvo, M.D.; Albert-Botella, L.; Pozo Giráldez, A.; et al. Circulating levels of calprotectin, a signature of neutrophil activation in prediction of severe respiratory failure in COVID-19 patients: A multicenter, prospective study (CalCov study). *Inflamm. Res.* **2022**, *71*, 57–67. [CrossRef] [PubMed]

biomedicines

Article

In Vitro Simulated Hemoperfusion on Seraph®-100 as a Promising Strategy to Counteract Sepsis

Antonio Lacquaniti [1], Antonella Smeriglio [2], Susanna Campo [1], Erminia La Camera [2], Giovanni Lanteri [2], Elena Giunta [3], Paolo Monardo [1,*,†] and Domenico Trombetta [2,*,†]

1 Nephrology and Dialysis Unit, Department of Internal Medicine, Papardo Hospital, 98158 Messina, Italy; ant.lacq@gmail.com (A.L.)
2 Department of Chemical, Biological, Pharmaceutical and Environmental Sciences, University of Messina, 98122 Messina, Italy; erminia.lacamera@unime.it (E.L.C.)
3 Pathology Unit, Papardo Hospital, 98158 Messina, Italy; elenagiunta@live.it
* Correspondence: pmonardo66@gmail.com (P.M.); domenico.trombetta@unime.it (D.T.); Tel.: +39-090-3996062 (P.M.); Fax: +39-090-3992337 (P.M.)
† These authors contributed equally to the paper.

Abstract: Blood purification represents a treatment option for sepsis, improving inflammation and the hyper-activated immune system. This study investigates the binding efficacy of Seraph®-100 against 10^8 CFU/mL of *Staphylococcus aureus* (*S. aureus*), *Pseudomonas aeruginosa* (*P. aeruginosa*), and *Escherichia coli* (*E. coli*) during a simulated hemoperfusion treatment. The fluorescence-activated cell sorting (FACS) technique was used to evaluate the bacteria reduction, whereas kinetic analysis and cultures revealed bacterial detection and counting at established time points. At the end of the experiment, the filter was cut at three different levels, obtaining suspensions for cultures and scanning electron microscopy (SEM) analyses. The FACS technique revealed a 78.77% reduction of the total bacterial load at the end of the treatment, with maximum filter sequestration occurring in the first 30 min of the treatment. Non-linear regression analysis of kinetic experiments ($T_{0–240\ min}$) highlighted a lower growth rate of *S. aureus* than the other two Gram bacteria, demonstrating a greater affinity without influencing a reduction rate of 99% for all three bacteria. The analyses of the suspension aliquots of the filter sections confirmed these data, revealing 1×10^8 CFU/mL, equal to the initial bacterial charge. Furthermore, the filter head adsorbed approximately 50% of bacteria, whereas the remaining amount was equally distributed between the body and the tail, as corroborated by SEM analysis. In conclusion, Seraph®-100 adsorbed 10^8 CFU/mL of *S. aureus*, *E. coli*, and *P. aeruginosa* during an in vitro simulated hemoperfusion session.

Keywords: acute kidney injury; adsorption; renal replacement therapy; sepsis; Seraph®-100

Citation: Lacquaniti, A.; Smeriglio, A.; Campo, S.; La Camera, E.; Lanteri, G.; Giunta, E.; Monardo, P.; Trombetta, D. In Vitro Simulated Hemoperfusion on Seraph®-100 as a Promising Strategy to Counteract Sepsis. *Biomedicines* **2024**, *12*, 575. https://doi.org/10.3390/biomedicines12030575

Academic Editors: Wen-Lin Su, Sheng-Kang Chiu, Chih-Hao Shen and Yi-Ting Chen

Received: 24 January 2024
Revised: 27 February 2024
Accepted: 29 February 2024
Published: 5 March 2024

1. Introduction

Sepsis, a complex and systemic disorder resulting from a dysregulated host response to an infection, leads to acute organ dysfunction and a high risk of death [1]. In 2020, about three million incident cases of sepsis were recorded in Europe [2], and the lung (64% of cases), followed by the abdomen (20%), bloodstream (15%), and renal and genitourinary tracts (14%), represented the most common sites of infection [3]. In particular, the Sepsis Occurrence in Acutely Ill Patients (SOAP) study reported a similar prevalence of Gram+ and Gram− bacterial infections among septic patients, with *Staphylococcus aureus*, *Pseudomonas* sp., and *Escherichia coli* being the most frequently identified organisms [4].

Behind epidemiological and clinical considerations, pharmacoeconomic implications are not negligible because they are related to the increased costs of sepsis when multidrug-resistant (MDR) bacteria occur [5,6]. In 2017, the World Health Assembly urged the World Health Organization member states to prioritize sepsis in their national health systems, recognizing sepsis as a global health priority [7]. Five years after this resolution, the challenge is always ongoing: facing the lack of standardized definitions or not applying them uniformly, reducing the delay of microbiological services to deliver blood culture results, and improving interdisciplinary collaboration. However, over the past decades, growing literature data have enhanced the timing of the diagnosis and treatment of sepsis, considering it not only an inflammatory disorder but highlighting an abnormal host response, triggering acute complications and organ failures [8,9].

In this perspective, a recent study defines sepsis as a syndromic entity, distinguishing "inflammopathic" or "coagulopathic" endotypes with the consequent personalization of treatment strategies [10]. Leaving aside this "research approach", screening for signs and symptoms of sepsis and septic shock facilitates earlier identification and intervention in clinical practice, aiming at the efficient and early removal of endotoxins and inflammatory factors and optimizing patient therapies and outcomes [11]. Antibiotics, the most effective weapon against bacterial infections, are becoming less effective, and new molecules are responsible for high costs. Therefore, there is an urgent need for alternative treatment options. In recent years, continuous blood purification therapies have represented a treatment option for sepsis, removing inflammatory mediators and acting on the hyper-activation of the innate and acquired immune systems, reducing the negative systemic response [12,13]. Whereas a single targeted cytokine removal was unsuccessful, the unselective removal of cytokines by hemoperfusion using adsorber systems, such as CytoSorb, improved the outcomes in septic patients [14,15]. However, contrasting data should be underlined by analyzing several studies and meta-analyses, not revealing a reduction in mortality by treatment with this adsorber filter [16]. Conflicting reports refer to the oXiris hemofilter, a hydrogel structure of the AN69 membrane coated with polyethyleneimine and heparin that adsorbs endotoxins and inflammatory cytokines [17]. Recent data-related oXiris filters have improved mortality, sequential organ failure assessment (SOFA) scores, and intensive care unit (ICU) stays [18]. In this context, encouraging data are emerging about the Seraph®-100 Microbind Affinity Filter (Exthera Medical, Martinez, CA, USA) [19]. Whereas the absorber filters reduced inflammatory response mediators, lipopolysaccharide, and endotoxins, Seraph®-100, based on ultrahigh-molecular-weight polyethylene beads with end point-attached heparin, removes bacteria, fungi, and viruses irreversibly from the bloodstream through binding with the immobilized heparin and miming the interaction with heparan sulfate on the cell surface (Figure 1) [20].

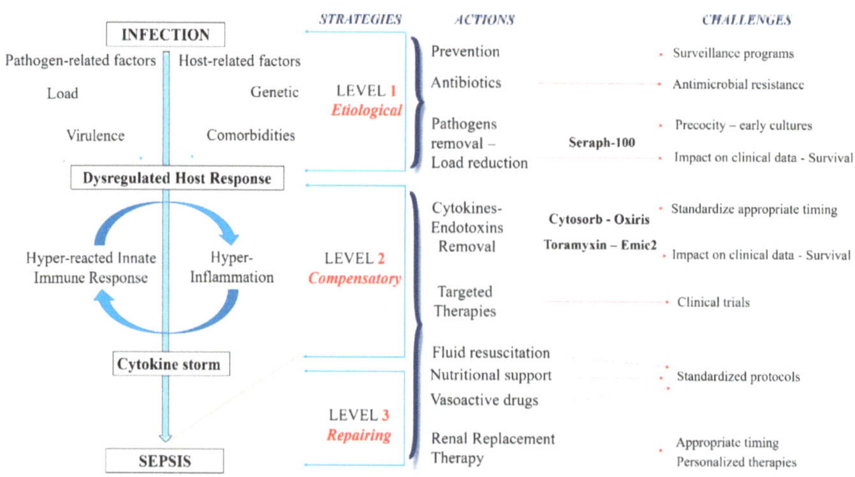

Figure 1. Strategies, actions, and challenges in septic patients.

To date, literature data refer to single-center experiences, often referring to the COVID-19 pandemic, complicated or not by sepsis, with positive effects on hemodynamic parameters and vasopressor requirements when this filter is applied [21,22]. Other results have been obtained in hemodialysis patients with catheter-related bacteremia, resulting in a faster resolution of bloodstream infections if treated with Seraph®-100 precociously within 66 h after the initial positive blood culture [23]. The correct timing for using this filter, whether applied alone or associated with continuous renal replacement therapy (CRRT), is an important topic. The preliminary results suggest "the sooner, the better", considering that applying the Seraph®-100 filter after 60 h from ICU admission or bacterial infection is associated with poor outcomes [24]. Several studies analyzed the performances of this filter in vitro, testing its binding activity to several pathogens with not negligible bias due to the use of miniaturized micro-columns pre-treated with infected solutions [25] or analyzing the effects after single passage through a column packed with heparinized beads in a laboratory setting, far from clinical practice [26]. Other considerations could be obtained by extrapolating data from in vitro studies analyzing the effects of Seraph®-100 on drug pharmacokinetics. Seraph®-100 did not influence the clearance of several antibiotics, except for aminoglycosides, including vancomycin, gentamicin, meropenem, and imipenem, with consequent positive clinical implications [27]. However, further studies should evaluate and confirm if a drug dose adjustment is required during Seraph®-100 use, avoiding sub-therapeutic antibiotic levels, which negatively influence the success of clinical trials, sepsis treatment, and overall survival. Starting from these assumptions, sepsis can be caused by multiple microorganisms, and this study investigates the binding efficacy of Seraph®-100 against three common sepsis pathogens, such as *Staphylococcus aureus* (Gram+), *Pseudomonas aeruginosa*, and *Escherichia coli* (Gram-), during a simulated hemoperfusion session. This preliminary in vitro study, simulating in all respects clinical practice, has the objective of clarifying the binding affinity of Gram+ and Gram- bacteria, present contextually as well as in the clinical condition of interest, and, at the same time, evaluating the filter ability to break down the simulated bacterial load to have a reference value of filter "binding ability" carrying out this clinical study on sepsis-affected patients.

For this purpose, a kinetic study was carried out, evaluating, for each time point, the bacterial counting by fluorescence-activated cell sorting (FACS) analysis and the detection of the number of colony-forming units (CFUs) by selective culture media. Finally, to verify the CFUs attached to the filter, the latter was divided into three sections, and a portion of the stationary phase was cultured on selective media for the three investigated bacterial strains and subjected to ultrastructure analysis by scanning electron microscopy (SEM).

2. Materials and Methods

2.1. Microbial Strains and Preparation of the Culture Suspension

The following bacterial strains from the University of Messina's in-house culture collection (Messina, Italy) were used for the preparation of the microbial suspension: *Staphylococcus aureus* ATCC 6538, *Escherichia coli* ATCC 10536, and *Pseudomonas aeruginosa* ATCC 9027. The overnight starter cultures were grown in Tryptic Soy Broth (TSB, Oxoid, CM0129) at 37 °C for 24 h and washed three times in filtered phosphate-buffered saline (PBS) by centrifugation at 3500× *g* for 10 min. The bacterial suspensions were then diluted to a density of approximately 10^8 CFU/mL, as spectrophotometrically recorded at 560 nm (Shimadzu UV-1601, Kyoto, Japan), and inoculated by syringe into a sterile bag containing 1000 mL of a 0.9% NaCl solution.

2.2. In Vitro Simulated Adsorption with Seraph®-100 and Experimental Design

A schematic overview of the experimental setup is given in Figure 2.

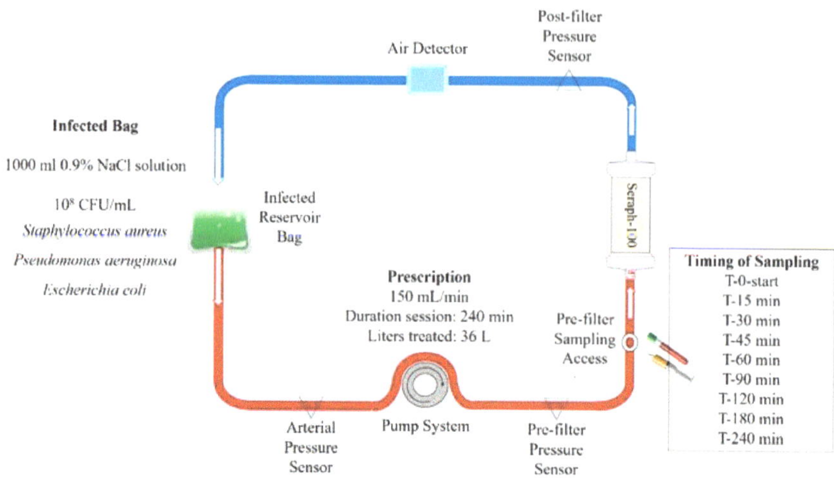

Figure 2. Schematic overview of the experimental set-up.

A Seraph®-100 disposable broad-spectrum absorbent device for extracorporeal blood purification (ExThera Medical, Martinez, CA, USA), consisting of ultra-high molecular weight polyethylene microspheres coated with heparin/heparan sulfate, was attached to a renal replacement therapy (RRT) device (Multifiltrate, Fresenius Medical Care AG & Co. KGaA, Bad Homburg, Germany), establishing a closed circuit. A flow rate of 150 mL/min was applied, and the Seraph®-100 was pre-treated with a 500 mL 0.9% NaCl solution, according to the manufacturer's instructions. After the filling of the entire circuit, determining a negligible extracorporeal volume of 100 mL, the reservoir bag containing the bacterial suspension (see previous section for specifications) was added to the closed circuit, starting the perfusion, which was then continued for 4 h (150 mL/min). The reservoir bag was periodically shaken during the procedure to ensure that the mixture of bacteria and samples was taken at the pre-filter level before the starting time (0 min) and at defined time points during the session (15, 30, 45, 60, 90, 120, 180, and 240 min). The experiment was repeated three times, and a microbiologist analyzed all samples in a blinded manner. A circuit with the same setup without the Seraph®-100 filter was used as a control.

2.3. *Bacterial Cell Counting*

2.3.1. Fluorescence-Activated Cell Sorting (FACS) Analysis

The FACS technique was used to quantify the three bacterial populations simultaneously. In the specific case, since the objective of this analysis was to count the CFUs present regardless of the specific bacterium present, to evaluate the reduction of the total bacterial load, only the light scattering technique, based on the deviation of the light beam based on the physical characteristics of the particles, was chosen. Two informative parameters were collected: forward scatter (FSC), which provides information on the size of the analyzed particles, and side scatter (SSC), to determine the graininess, roughness, or nucleus/cytoplasm ratio. The graph that can be obtained from this simple analysis is a one-dimensional cytogram or a two-dimensional dot-plot, where each dot represents a single cell detected and analyzed. This analysis, conducted against a specific reaction blank, allows for the evaluation of the reduction of the charge during the treatment, setting the bacterial count of the original bacterial solution to 100% to calculate the percent removal. FACS analyses were performed using a clinical flow cytometry system (BD FACS Canto™, Milan, Italy) according to the manufacturer's instructions.

2.3.2. Detection of Bacterial Strains by Selective Culture Media

For bacterial strain (*S. aureus, E. coli,* and *P. aeruginosa*) detection and counting, 50 µL of each time point of the kinetic experiment (0, 15, 30, 45, 60, 90, 120, 180, and 240 min) were sown in selective culture media: Baird Parker agar (BP), Tergitol TTC agar (TTC), and *Pseudomonas* CN agar (CN), respectively. Cultures were incubated for 48 h at 37 °C.

2.3.3. Evaluation of the Bacteria-Binding Capacity

To evaluate the bacteria-binding capacity of the filter and its possible selectivity for a specific bacterial strain, at the end of the experiment (240 min), the reservoir bag, the filter, and the waste bag were analyzed in triplicate by a blinded microbiologist. Specifically, 50 µL of the reservoir and waste bag content was plated directly into the selective agar culture media (BP, TTC, and CN), whereas the filter was cut under sterile conditions into three sections: head (H), body (B), and tail (T). The content of the three filter sections was recovered and used to prepare suitable suspensions (5 mg/mL), which were also seeded (50 µL) in the selective agar culture media (BP, TTC, and CN). Cultures were incubated for 48 h at 37 °C. Finally, to verify that the bacteria had effectively adhered to the surface of the filter stationary phase, samples of the three filter sections were also investigated through SEM analysis with a Zeiss EVO MA10 (Carl Zeiss S.p.A., Milan, Italy) at an acceleration voltage of 20 kV. Filter stationary phase H, B, and T samples (1 mg) were fixed in 70% ethanol for 48 h and dehydrated through an ethanol series (90% and 100%, 1 h each). Samples were mounted on stubs (SEM-PT-F-12), covered by conductive adhesive tables, and left at 28 °C for 12 h, avoiding the critical drying point, before being covered with 20 nm gold palladium.

2.4. *Statistical Analyses*

Three independent experiments in triplicate ($n = 3$) were carried out. The results, expressed as n. CFU as a function of time (min), were analyzed using a non-linear regression approach based on the three-parameter sigmoid equation followed by the Shapiro–Wilk test using SigmaPlot 12.0 software (Systat Software Inc., San Jose, CA, USA). Data were considered statistically significant for $p < 0.05$.

3. Results

All experiments were performed under stable conditions without technical problems, and the cartridge was perfused without interruptions.

To evaluate the filter's ability to break down the bacterial load and the specific binding affinity of the three bacterial strains towards the heparin-coated beads, an in vitro simulated perfusion with three reference strains, one Gram+ (*S. aureus*) and two Gram- bacteria (*P. aeruginosa* and *E. coli*), each 10^8 CFU/mL, suspended in a 0.9% NaCl solution, was carried out by Seraph®-100 according to the manufacturer's instructions.

The analyses of kinetics points ($T_{0\text{–}240\ min}$) by the FACS technique allowed for calculating the total reduction of the bacterial load at the end of the treatment as being equal to 78.77% (7.88×10^7 CFU/mL). The two-dimensional dot-plot and one-dimensional cytogram of bacterial cell counting obtained by FACS analyses at the starting point (T_0, Figure 3A and Figure 3B, respectively) and at the end of the simulated in vitro perfusion session ($T_{240\ min}$, Figure 3C and Figure 3D, respectively) with Seraph®-100 are depicted in Figure 3.

The load reduction recorded at the various times also allows for clarifying, with absolute certainty, that the maximum filter sequestration of the bacterial strains already occurs in the first 30 min of treatment, as confirmed by the plate count carried out on selective media for the three investigated bacterial strains (Figure 4).

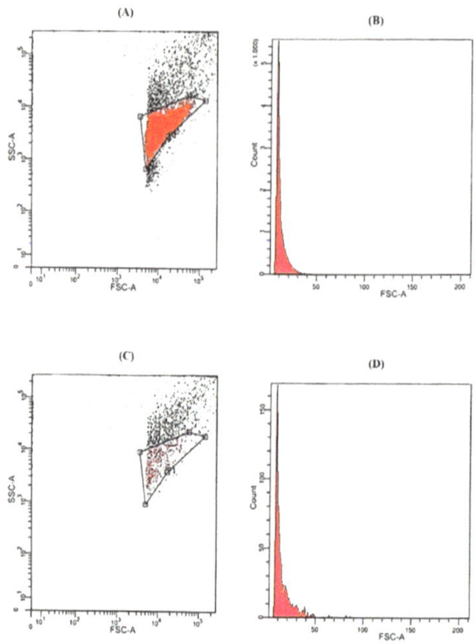

Figure 3. Two-dimensional dot-plot and one-dimensional cytogram of bacterial cell counting obtained by FACS analyses at the starting point (T_0, (**A**) and (**B**), respectively) and at the end of the simulated in vitro perfusion session ($T_{240\ min}$, (**C**) and (**D**), respectively) with Seraph®-100.

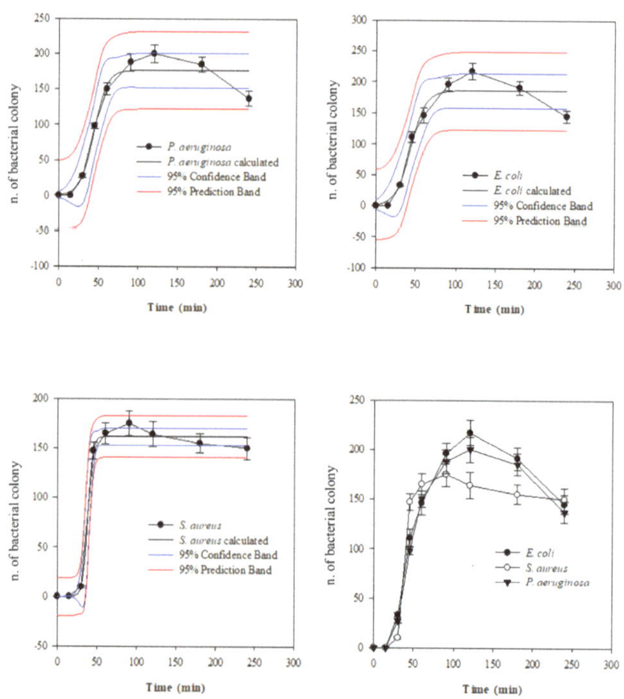

Figure 4. Non-linear regression analysis of kinetic experiments ($T_{0-240\ min}$). Results, which represent the average of three independent experiments in triplicate (n = 3), were expressed as the n. of colony-forming units (CFUs) with respect to time (min), as detected by selective culture media.

The results, expressed as n. CFU as a function of time (min), were analyzed using a non-linear regression approach based on the three-parameter sigmoid equation reported below:

$$y = \frac{a}{1 + e^{-\left(\frac{x - x0}{b}\right)}}$$

where

a = upper asymptote
b = slope (growth rate)
$x0$ = crossover point (time of maximum growth)

In Figure 4, beyond the growth curve of each investigated bacterial strain (*P. aeruginosa*; *E. coli*; *S. aureus*), confidence (blue lines) and predicted bands (solid black lines) were depicted. The first ones were used to represent the uncertainty in estimating a curve based on limited or noisy data, while the prediction band was used to represent the uncertainty in the value of a new data point on the curve subject to noise. As can be seen from the three panels of Figure 4, the growth curves fall perfectly within the confidence bands and follow a trend that is almost superimposable to that of the predicted band. This demonstrates the reliability of the recorded data.

Table 1 shows the comparison of the parameters used for the non-linear regression analysis for each investigated bacterial strain and the Shapiro–Wilk normality test, used to evaluate the statistical significance of the results recorded for each kinetic data point depicted in Figure 4 for each investigated bacterial strain.

Table 1. Comparison of the sigmoid parameters of the three investigated bacterial strains.

Sigmoid Parameters	*E. coli*	*S. aureus*	*P. aureginosa*
a	185.856	161.812	176.860
b	9.386	2.986	8.227
X0	43.319	38.135	43.799
	Normality Test (Shapiro–Wilk) P/F *		
	P	P	P

* P = pass; F = fail. The Shapiro–Wilk normality test was used to evaluate the statistical significance ($p < 0.05$) of each data point of the kinetic curve of each investigated bacterial strain.

As can be seen from Figure 4, in which it is possible to observe an overlap of the three growth curves of the three investigated bacterial strains, and from Table 1, the slope or growth rate of *S. aureus* (Gram+) is significantly lower than the other two investigated bacterial strains, both Gram−. This demonstrates a greater affinity of Gram+, and in this case of *S. aureus*, for Seraph®-100, which translates into a greater sequestration rate of this bacterial strain during the simulated in vitro perfusion procedure. However, it is worth underlining that the efficiency of reducing the bacterial load of the three investigated strains individually by counting them on selective media reveals a very efficient ability of the filter to sequester the three investigated bacterial strains with reductions equal to 99.96%, 99.85%, and 99.87%, i.e., leading to a final bacterial load of 4×10^4 CFU/mL, 1.5×10^5 CFU/mL, and 1.3×10^5 CFU/mL vs. the starting bacterial load of 1×10^8 CFU/mL for *S. aureus*, *E. coli*, and *P. aeruginosa*, respectively. These data, recorded on the samples ($T_{0\text{–}240\ \text{min}}$) taken during the kinetic experiment, were further validated by the analyses conducted on the filter sections: H, B, and T. Indeed, to experimentally verify whether the bacteria remained bonded to the heparin/heparan sulfate-coated beads, the filter was sectioned into three parts, and suspensions of representative aliquots of the stationary phase of each section were prepared. Knowing the weight of the entire stationary phase, the n. CFU was calculated (H + B + T), with an average value of about 1×10^8 for *S. aureus*, *E. coli*, and *P. aeruginosa*, respectively, according to the above results. Furthermore, this experiment also allowed for verifying the distribution of bacteria within the stationary phase of the filter, which are located approximately 50% in filter H and the remaining 50% equally between filters B and T. The binding affinity of the bacterial strains to heparin/heparan sulfate-coated beads was also corroborated by SEM analysis, as depicted in Figure 5.

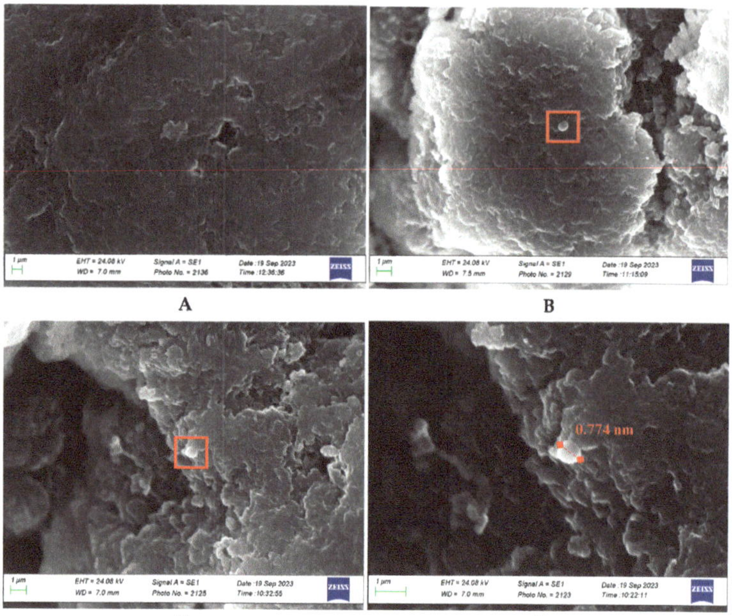

Figure 5. Representative SEM micrograph of the blank stationary phase (heparin/heparan sulfate-coated beads) of Seraph®-100 (**A**) in comparison with the Seraph®-100 stationary phase at the end of perfusion time (4 h). (**B**,**C**) show some bacteria (red rectangle) attached to the filter stationary phase; (**D**) shows a magnification of (**C**), in which the bacterial size is highlighted.

4. Discussion

This study demonstrated that Seraph®-100 adsorbed three bacteria from a super-infected solution during an in vitro simulated hemoperfusion session, suggesting a new era in treating bloodstream infections caused by several pathogens, often inducing sepsis. A bacterial load is related to the severity of sepsis and increased mortality [28–30]. Seraph®-100 removed 10^8 CFU/mL of *S. aureus*, *E. coli*, and *P. aeruginosa* contemporary inoculated into a 1000 mL 0.9% NaCl solution, a concentration higher than 100 CFU/mL commonly observed in adult patients with bacteremia [31].

Interestingly, we noted a greater affinity of Seraph®-100 towards the Gram+ *S. aureus* when comparing the kinetic data of Gram- pathogens such as *E. coli* and *P. aeruginosa*.

Electrical and chemical bonds mediate this adsorption through coated microspheres with immobilized heparin, which mimics the heparan sulfate (HS), a type of sulfated glycosaminoglycan of the endothelial glycocalyx, which is targeted by several different pathogens as an initial attachment site during their pathogenesis [32]. This allows for the formation of a negative electrostatic barrier that separates negatively charged blood components, such as red blood cells, and, at the same time, catches elements with opposite electric charges and through specific ligands, often localized in the bacterial glycocalyx (Figure 6).

These challenges between bacteria, endothelium, and immune cells occurred similarly within Seraph®-100, and probably the thinner peptidoglycan mesh surrounding Gram- cells, compared to the thicker layers of peptidoglycan and glycocalyx observed in *S. aureus*, could support the different absorption underlined by our results. At the same time, the similar chemical structure of peptidoglycan in Gram+ and Gram− bacteria could also explain the non-selective effects of Seraph®-100 on bacteria, viruses, and fungi [33]. However, these bonds are not mediated exclusively by HS, as observed in *P. aeruginosa*, which could interact with the cells through specific poly-cationic ligands [34] or in *E. coli*, whose attachment to heparin-coated polyethylene beads depends on fimbrial adhesin expressions [35,36].

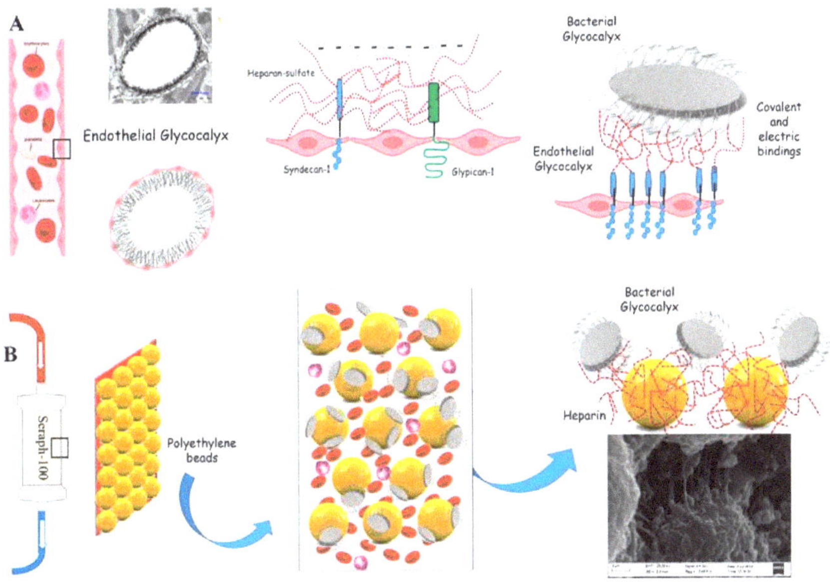

Figure 6. Adsorption processes in Seraph®-100 heparin-coated microspheres, which mimic the heparin sulfate of the endothelial glycocalyx. (**A**) The endothelial glycocalyx and bindings of bacteria. (**B**) Seraph structure miming the endothelial glycocalyx.

Recently, Seffer revealed the ability of this filter to adsorb *S. aureus* using miniaturized and engineered Seraph®-100 adsorbers through pre-treated micro-columns with an infected 0.9% NaCl solution and infected human plasma [25].

Conversely, our data refer to an in vitro model applying to a circuit of hemoperfusion, the commercialized Seraph®-100 filter, which was stressed by the crossing of 36 L of a super-infected solution, miming clinical practice, and not a laboratory test. Our analyses revealed specific kinetic adsorption models for each bacterium and assessed the reduction of more than 90% of bacterial CFUs, not only for *S. aureus* but also for *E. coli* and *P. aeruginosa*. These data strengthen the already-published ones, highlighting the ability of this adsorber to remove several pathogens [20].

However, whereas these studies analyzed a single-pass removal of a single bacterium through the filter or its miniaturized columns, we tested the commercialized filter, revealing that the contemporary presence of the three bacteria did not influence their single adsorption, considering Seraph®-100 potentially effective also in patients with superinfections as often observed in clinical practice.

Behind these kinetic data, this study evaluated the adsorption properties of Seraph®-100, analyzing the distribution of the attached bacteria throughout the filter, underlying that the head was the principal site of adsorption, with less attachment recorded in the body and the filter tail. This distribution, confirmed by SEM analyses, suggests an early saturation of the first area of Seraph®-100 and, associated with the peak of bacterial CFU reduction after 30 min, hypothesizes a time-dependent adsorption mechanism and a strong bond between bacteria and heparin-coated beds. This latter concept is essential for the safe use of Seraph®-100, explaining the absent back-release of attached pathogens from the filter to the bloodstream after their irreversible adsorption [23].

These data allow for considering Seraph®-100 as different from other blood filters often used in septic patients, acting on a precocious etiological level of the sepsis cascade and preventing it through the direct removal of the infective agents from the bloodstream. Conversely, other adsorber devices and hemofilters applied to hemoperfusion or renal replacement therapies reduced cytokines or endotoxins. According to these results, a combined therapy could be a new approach in bacteremic patients, based on a precocious treatment with Seraph®-100 associated in series with highly permeable/high cut-off membranes for the prevention of sepsis, eliminating pathogens from the bloodstream, and removing inflammatory mediators, with the possibility of breaking the "cytokine storm". Moreover, dead bacterial cells or fragments of their cell walls, as observed during the antibiotic treatment, may induce inflammation and harmful dysregulated host responses, whose elimination from the bloodstream could increase host tolerance and bactericidal mechanisms by the innate and acquired immune systems [37].

However, further studies should confirm in vivo the efficacy of Seraph®-100 in infected patients, trying to solve significant unmet needs, such as the timing of intervention. Clinical trials are required to assess clinically relevant endpoints such as the incidence of sepsis, hospital length of stay, and mortality. The challenge is also to test this filter in series with other devices, trying to hit more pathways that are over-expressed in a critically infected patient.

At the same time, antimicrobial dose optimization in patients undergoing CRRT is challenging, and despite its growing use in critically ill patients, the paucity of pharmacokinetic data during CRRT limits evidence-based antibiotic dosing recommendations for novel agents. Further data are required to evaluate the efficacy and clinical application of adsorber filters [38,39].

In vitro studies revealed that Seraph®-100 did not remove several antibiotics, but further in vivo studies should evaluate and confirm if a drug dose adjustment is required, considering its non-selective adsorption process and avoiding sub-therapeutic antibiotic levels, which negatively influence the success of sepsis treatment and overall survival.

This study has several limitations. The experimental model was based on analyses conducted in a hemoperfusion circuit associated with the commercialized filter but analyzed the adsorption effects of inoculated pathogens in a saline solution, not comparable to the human whole blood. For this reason, potential interactions between bacteria, blood components, such as cells and proteins, and Seraph®-100 could not be investigated. Furthermore, the absence of concomitant antibiotic therapy should be underlined.

However, this study simulated clinical practice, demonstrating the effectiveness of this treatment.

5. Conclusions

Seraph®-100 represents an efficient device to remove, through adsorption, *S. aureus*, *E. coli*, and *P. aeruginosa*, suggesting its precocious use during bacteremia or superinfection. Further studies should confirm in vivo the efficacy of Seraph®-100 in infected patients, highlighting that the complexity of sepsis cannot be faced and solved only through a single device but by applying different therapeutic strategies, for example, the combined use of Seraph®-100 and hemofilters. A multidisciplinary team should design clinical trials to overcome the limitations of this study.

Author Contributions: Conceptualization, A.L., A.S., P.M., and D.T.; methodology, A.L., A.S., P.M., and D.T.; validation, S.C., E.L.C., and G.L.; formal analysis, E.G., E.L.C., and G.L.; investigation, S.C., E.G., E.L.C., and G.L.; data curation, S.C., E.L.C., and G.L.; writing—original draft preparation, A.L. and A.S.; writing—review and editing, A.L., A.S., and S.C.; visualization, P.M. and D.T.; supervision, P.M. and D.T. All authors have read and agreed to the published version of the manuscript.

Funding: This research received no external funding.

Institutional Review Board Statement: Not applicable for this study, not involving humans or animals.

Informed Consent Statement: Not applicable for this study, not involving humans.

Data Availability Statement: The data underlying this article will be shared on reasonable request to the corresponding author.

Acknowledgments: The authors acknowledge Francesco Fucile, Caterina Ragno, and Carmelo Saterno for preparing and managing the hemoperfusion circuit and performing the samplings.

Conflicts of Interest: The authors declare no conflicts of interest.

References

1. Nordio, M.; Reboldi, G.; Di Napoli, A.; Quintaliani, G.; Alberici, F.; Postorino, M.; Aucella, F.; Messa, P.; Brunori, G. Italian Society of Nephrology COVID-19 Research Group: Risk factors and action thresholds for the novel coronavirus pandemic. Insights from the Italian Society of Nephrology COVID-19 Survey. *J. Nephrol.* **2021**, *34*, 325–335. [CrossRef]
2. Rudd, K.E.; Johnson, S.C.; Agesa, K.M.; Shackelford, K.A.; Tsoi, D.; Kievlan, D.R.; Colombara, D.V.; Ikuta, K.S.; Kissoon, N.; Finfer, S.; et al. Global, regional, and national sepsis incidence and mortality, 1990–2017: Analysis for the Global Burden of Disease Study. *Lancet* **2020**, *395*, 200–211. [CrossRef] [PubMed]
3. Vincent, J.L.; Rello, J.; Marshall, J.; Silva, E.; Anzueto, A.; Martin, C.D.; Moreno, R.; Lipman, J.; Gomersall, C.; Sakr, Y.; et al. International study of the prevalence and outcomes of infection in intensive care units. *JAMA* **2009**, *302*, 2323–2329. [CrossRef] [PubMed]
4. Vincent, J.L.; Sakr, Y.; Sprung, C.L.; Ranieri, V.M.; Reinhart, K.; Gerlach, H.; Moreno, R.; Carlet, J.; Le Gall, J.-R.; Payen, D.; et al. Sepsis in European intensive care units: Results of the SOAP study. *Crit. Care Med.* **2006**, *34*, 344–353. [CrossRef] [PubMed]
5. Nelson, R.E.; Hatfield, K.M.; Wolford, H.; Samore, M.H.; Scott, R.D.; Reddy, S.C.; Olubajo, B.; Paul, P.; Jernigan, J.A.; Baggs, J. National Estimates of Healthcare Costs Associated with Multidrug-Resistant Bacterial Infections Among Hospitalized Patients in the United States. *Clin. Infect. Dis.* **2021**, *72*, S17–S26. [CrossRef] [PubMed]
6. Hernandez-Pastor, L.; Geurtsen, J.; Baugh, B.; El Khoury, A.C.; Kalu, N.; Krishnarajah, G.; Gauthier-Loiselle, M.; Bungay, R.; Cloutier, M.; Saade, E. Economic burden of invasive Escherichia coli disease among older adult patients treated in hospitals in the United States. *J. Manag. Care Spec. Pharm.* **2023**, *29*, 873–883. [CrossRef]
7. Reinhart, K.; Daniels, R.; Kissoon, N.; Machado, F.R.; Schachter, R.D.; Finfer, S. Recognizing Sepsis as a Global Health Priority—A WHO Resolution. *N. Engl. J. Med.* **2017**, *377*, 414–417. [CrossRef]
8. Evans, L.; Rhodes, A.; Alhazzani, W.; Antonelli, M.; Coopersmith, C.M.; French, C.; Machado, F.R.; Mcintyre, L.; Ostermann, M.; Prescott, H.C.; et al. Executive summary: Surviving sepsis campaign: International guidelines for the management of sepsis and septic shock 2021. *Critic. Care Med.* **2021**, *49*, 1974–1982. [CrossRef]
9. Lacquaniti, A.; Ceresa, F.; Campo, S.; Barbera, G.; Caruso, D.; Palazzo, E.; Patanè, F.; Monardo, P. Acute Kidney Injury and Sepsis after Cardiac Surgery: The Roles of Tissue Inhibitor Metalloproteinase-2, Insulin-like Growth Factor Binding Protein-7, and Mid-Regional Pro-Adrenomedullin. *J. Clin. Med.* **2023**, *12*, 5193. [CrossRef]
10. Balch, J.A.; Chen, U.I.; Liesenfeld, O.; Starostik, P.; Loftus, T.J.; Efron, P.A.; Brakenridge, S.C.; Sweeney, T.E.; Moldawer, L.L. Defining critical illness using immunological endotypes in patients with and without sepsis: A cohort study. *Crit. Care* **2023**, *27*, 292. [CrossRef]
11. Levy, M.M.; Dellinger, R.P.; Townsend, S.R.; Linde-Zwirble, W.T.; Marshall, J.C.; Bion, J.; Schorr, C.; Artigas, A.; Ramsay, G.; Beale, R.; et al. The Surviving Sepsis Campaign: Results of an international guideline-based performance improvement program targeting severe sepsis. *Crit. Care Med.* **2010**, *38*, 367–374. [CrossRef] [PubMed]
12. Girardot, T.; Schneider, A.; Rimmelé, T. Blood purification techniques for sepsis and septic AKI. *Semin. Nephrol.* **2019**, *39*, 505–514. [CrossRef] [PubMed]
13. Campo, S.; Lacquaniti, A.; Trombetta, D.; Smeriglio, A.; Monardo, P. Immune System Dysfunction and Inflammation in Hemodialysis Patients: Two Sides of the Same Coin. *J. Clin. Med.* **2022**, *11*, 3759. [CrossRef] [PubMed]
14. Ricci, Z.; Romagnoli, S.; Reis, T.; Bellomo, R.; Ronco, C. Hemoperfusion in the intensive care unit. *Intensive Care Med.* **2022**, *48*, 1397–1408. [CrossRef] [PubMed]
15. Harm, S.; Schildböck, C.; Hartmann, J. Cytokine removal in extracorporeal blood purification: An in vitro study. *Blood Purif.* **2020**, *49*, 33–43. [CrossRef]
16. Becker, S.; Lang, H.; Vollmer Barbosa, C.; Tian, Z.; Melk, A.; Schmidt, B.M.W. Efficacy of CytoSorb: A systematic review and meta-analysis. *Crit. Care* **2023**, *27*, 215. [CrossRef]
17. Broman, M.E.; Hansson, F.; Vincent, J.L.; Bodelsson, M. Endotoxin and cytokine reducing properties of the oXiris membrane in patients with septic shock: A randomized crossover double-blind study. *PLoS ONE* **2019**, *14*, e0220444. [CrossRef]
18. Wang, G.; He, Y.; Guo, Q.; Zhao, Y.; He, J.; Chen, Y.; Chen, W.; Zhou, Y.; Peng, Z.; Deng, K.; et al. Continuous renal replacement therapy with the adsorptive oXiris filter may be associated with the lower 28-day mortality in sepsis: A systematic review and meta-analysis. *Crit. Care* **2023**, *27*, 275. [CrossRef]
19. Schmidt, J.; Eden, G.; Seffer, M.; Winkler, M.; Kielstein, J. In vitro elimination of anti-infective drugs by the Seraph®100 Microbind affinity blood filter. *Clin. Kidney J.* **2020**, *13*, 421–424. [CrossRef]

20. Seffer, M.T.; Cottam, D.; Forni, L.G.; Kielstein, J.T. Heparin 2.0: A new approach to the infection crisis. *Blood Purif.* **2021**, *50*, 28–34. [CrossRef]
21. Stoffel, S.; Boster, J.; Jarrett, Z.; Rosas, M.; Kalra, A.; Nugyen, M.; Morris, M.; Walter, R. Single-Center Experience with the Seraph-100® Microbind® Affinity Blood Filter in Patients with SARS-CoV-2 Infection and Septic Shock at a Military Treatment Facility. *Mil. Med.* **2023**, *188*, e2670–e2674. [CrossRef] [PubMed]
22. Premuzic, V.; Situm, I.; Lovric, D.; Erceg, A.; Karmelic, D.; Mogus, M.; Jurjevic, M.; Nedeljkovic, V.; Mazar, M.; Mihaljevic, S.; et al. Sequential Extracorporeal Blood Purification Is Associated with Prolonged Survival among ICU Patients with COVID-19 and Confirmed Bacterial Superinfection. *Blood Purif.* **2023**, *52*, 642–651. [CrossRef] [PubMed]
23. Eden, G.; Schmidt, J.J.; Büttner, S.; Kümpers, P.; Hafer, C.; Rovas, A.; Koch, B.F.; Schmidt, B.M.W.; Kielstein, J.T. Safety and efficacy of the Seraph® 100 Microbind® Affinity Blood Filter to remove bacteria from the blood stream: Results of the first in human study. *Crit. Care* **2022**, *26*, 181. [CrossRef] [PubMed]
24. Schmidt, J.J.; Borchina, D.N.; Van't Klooster, M.; Bulhan-Soki, K.; Okioma, R.; Herbst, L.; Rodríguez, D.S.; Premužić, V.; Büttner, S.; Bader, B.; et al. Interim analysis of the COSA (COVID-19 patients treated with the Seraph® 100 Microbind® Affinity filter) registry. *Nephrol. Dial. Transplant.* **2022**, *37*, 673–680. [CrossRef]
25. Seffer, M.T.; Weinert, M.; Molinari, G.; Rohde, M.; Gröbe, L.; Kielstein, J.T.; Engelmann, S. Staphylococcus aureus binding to Seraph® 100 Microbind® Affinity Filter: Effects of surface protein expression and treatment duration. *PLoS ONE* **2023**, *18*, e0283304. [CrossRef]
26. Mattsby-Baltzer, I.; Bergstrom, T.; McCrea, K.; Ward, R.; Adolfsson, L.; Larm, O. Affinity Apheresis for Treatment of Bacteremia caused by Staphylococcus aureus and/or Methicillin-Resistant S. Aureus (MRSA). *J. Microbiol. Biotechnol.* **2011**, *21*, 659–664. [CrossRef]
27. Selig, D.J.; Reed, T.; Chung, K.K.; Kress, A.T.; Stewart, I.J.; De Luca, J.P. Hemoperfusion with Seraph 100 Microbind Affinity Blood Filter Unlikely to Require Increased Antibiotic Dosing. A Simulations Study Using a Pharmacokinetic/Pharmacodynamic Approach. *Blood Purif.* **2023**, *52*, 25–31.
28. Hackett, S.J.; Guiver, M.; Marsh, J.; Sills, J.A.; Thomson, A.P.; Kaczmarski, E.B.; Hart, C.A. Meningococcal bacterial DNA load at presentation correlates with disease severity. *Arch. Dis. Child.* **2022**, *86*, 44–46. [CrossRef]
29. Rello, J.; Lisboa, T.; Lujan, M.; Gallego, M.; Kee, C.; Kay, I.; Lopez, D.; Waterer, G.W. DNA-Neumococo Study Group. Severity of pneumococcal pneumonia associated with genomic bacterial load. *Chest* **2009**, *136*, 832–840. [CrossRef]
30. Shao, Z.; Zhu, J.; Wei, Y.; Jin, J.; Zheng, Y.; Liu, J.; Zhang, R.; Sun, R.; Hu, B. Pathogen load and species monitored by droplet digital PCR in patients with bloodstream infections: A prospective case series study. *BMC Infect. Dis.* **2022**, *22*, 771. [CrossRef] [PubMed]
31. Huang, T.H.; Tzeng, Y.L.; Dickson, R.M. FAST. Rapid determinations of antibiotic susceptibility phenotypes using label-free cytometry. *Cytometry A* **2018**, *93*, 639–648. [CrossRef]
32. Bartlett, A.H.; Park, P.W. Heparan Sulfate Proteoglycans in Infection. *Glycans Dis. Ther.* **2011**, *19*, 31–62.
33. Silhavy, T.J.; Kahne, D.; Walker, S. The bacterial cell envelope. *Cold Spring Harb. Perspect. Biol.* **2010**, *2*, a000414. [CrossRef]
34. Sanchez, H.; Hopkins, D.; Demirdjian, S.; Gutierrez, C.; O'Toole, G.A.; Neelamegham, S.; Berwin, B. Identification of cell-surface glycans that mediate motility-dependent binding and internalization of Pseudomonas aeruginosa by phagocytes. *Mol. Immunol.* **2021**, *131*, 68–77. [CrossRef] [PubMed]
35. Sahly, J.; Navon-Venezia, S.; Roesler, L.; Hay, A.; Carmeli, Y.; Podschun, R.; Hennequin, C.; Forestier, C.; Ofek, I. Extended Spectrum B-lactamase Production is Associated with an Increase in Cell Invasion and Expression of Fimbrial Adhesins in Klebsiella pneumoniae. *Antimicrob. Agents Chemother.* **2008**, *52*, 3029–3034. [CrossRef] [PubMed]
36. McCrea, K.; Ward, R.; LaRosa, S.P. Removal of Carbapenem-Resistant Enterobacteriaceae (CRE) from blood by heparin-functional hemoperfusion media. *PLoS ONE* **2014**, *9*, e114242. [CrossRef] [PubMed]
37. Minasyan, H. Sepsis: Mechanisms of bacterial injury to the patient. *Scand. J. Trauma. Resusc. Emerg. Med.* **2019**, *27*, 19. [CrossRef] [PubMed]
38. Harm, S.; Falkenhagen, D.; Hartmann, J. Pore size—A key property for selective toxin removal in blood purification. *Int. J. Artif. Organs* **2014**, *37*, 668–678. [CrossRef] [PubMed]
39. König, C.; Röhr, A.C.; Frey, O.R.; Brinkmann, A.; Roberts, J.A.; Wichmann, D.; Braune, S.; Kluge, S.; Nierhaus, A. In vitro removal of anti-infective agents by a novel cytokine adsorbent system. *Int. J. Artif. Organs* **2019**, *42*, 57–64. [CrossRef]

biomedicines

Article

The Neutrophil/Lymphocyte Ratio and Outcomes in Hospitalized Patients with Community-Acquired Pneumonia: A Retrospective Cohort Study

Aysun Tekin [1,*,†], Felix W. Wireko [2,†], Ognjen Gajic [2] and Yewande E. Odeyemi [2]

1 Division of Nephrology and Hypertension, Department of Internal Medicine, Mayo Clinic, Rochester, MN 55905, USA

2 Division of Pulmonary and Critical Care Medicine, Mayo Clinic, Rochester, MN 55905, USA; wireko.felix@mayo.edu (F.W.W.); gajic.ognjen@mayo.edu (O.G.); odeyemi.yewande@mayo.edu (Y.E.O.)

* Correspondence: tekin.aysun@mayo.edu; Tel.: +1-507-255-1162

† These authors contributed equally to this work.

Abstract: We aimed to assess the prognostic role of the neutrophil/lymphocyte ratio (NLR) in community-acquired pneumonia (CAP) via a single-center retrospective cohort of hospitalized adult patients from 1/2009 to 12/2019. Patients were dichotomized into lower NLR (≤12) and higher NLR (>12). The primary outcome was mortality. ICU admission and hospital- and ICU-free days were secondary outcomes. The pneumonia severity index (PSI) and the NLR's ability to predict outcomes was also tested. An NLR ≤12 was observed in 2513 (62.2%) patients and >12 in 1526 (37.8%). After adjusting for PSI, the NLR was not associated with hospital mortality (odds ratio [OR] 1.115; 95% confidence interval [CI] 0.774, 1.606; $p = 0.559$), but it was associated with a higher risk of ICU admission (OR 1.405; 95% CI 1.216, 1.624; $p < 0.001$). The PSI demonstrated acceptable discrimination for mortality (area under the receiver operating characteristic curve [AUC] 0.78; 95% CI 0.75, 0.82) which was not improved by adding the NLR (AUC 0.78; 95% CI 0.75, 0.82, $p = 0.4476$). The PSI's performance in predicting ICU admission was also acceptable (AUC 0.75; 95% CI 0.74, 0.77) and improved by including the NLR (AUC 0.76, 95% CI 0.74, 0.77, $p = 0.008$), although with limited clinical significance. The NLR was not superior to the PSI for predicting mortality in hospitalized CAP patients.

Keywords: community-acquired pneumonia; CAP; neutrophil/lymphocyte ratio; NLR; mortality; prognostication

Citation: Tekin, A.; Wireko, F.W.; Gajic, O.; Odeyemi, Y.E. The Neutrophil/Lymphocyte Ratio and Outcomes in Hospitalized Patients with Community-Acquired Pneumonia: A Retrospective Cohort Study. *Biomedicines* **2024**, *12*, 260. https://doi.org/10.3390/biomedicines12020260

Academic Editors: Wen-Lin Su, Sheng-Kang Chiu, Chih-Hao Shen and Yi-Ting Chen

Received: 31 December 2023
Revised: 18 January 2024
Accepted: 22 January 2024
Published: 24 January 2024

1. Introduction

Community-acquired pneumonia (CAP) is an acute infection of the lung parenchyma and one of the leading causes of hospitalization and mortality in the United States [1], with more than 1.5 million annual hospitalizations and about 100,000 deaths [2]. Risk stratification and prognostication provide useful information on the disease trajectory and guide management [3]. Although several prognostic tools for CAP have been evaluated, the pneumonia severity index (PSI), Confusion, Uremia, Respiratory rate, Blood pressure, being 65 years of age and older (CURB-65), and the ATS/IDSA are the most validated and remain commonly used in CAP, with satisfactory outcomes [4–6]. Evaluations of the prognostic value of biomarkers like NT-pro-BNP [7], C-reactive protein, and procalcitonin have been less satisfactory when used individually [8]. The neutrophil/lymphocyte ratio (NLR) (the ratio of absolute neutrophil count to absolute lymphocyte count), is an easily measurable index that is receiving growing interest as a useful prognostic biomarker in several stressful conditions including infections. In CAP, multiple studies have shown conflicting results on the prognostic value of the NLR alone or in addition to clinical severity scores [7,9–12]. The addition of the NLR to the PSI and CURB-65 in one study did not improve the prediction of mortality [12]. More recently, a systematic review demonstrated a comparable prognostic value of the NLR to the PSI, CURB-65, CRP, procalcitonin, neutrophil count, lymphocyte count, and white blood cell count [13]. However, only a few studies in the systematic review confirmed an association between the NLR and adverse outcomes utilizing a multivariate analysis, thus resulting in limited consideration of confounding factors on mortality. Also, most studies focused on mortality only as an outcome, with minimal evaluation of other clinical outcomes including need for invasive and non-invasive mechanical support and admission to the intensive care unit (ICU). Further, most studies were limited by sample size.

The aim of this study was to evaluate the prognostic value of the NLR in predicting the clinical outcomes of patients who are hospitalized with CAP.

2. Materials and Methods

The institutional review board (IRB) at the Mayo Clinic approved this research as low-risk, denoted by the IRB number 17-011140, with a waiver of the requirement for written informed consent.

2.1. Design, Setting, and Participants

This study had a retrospective cohort design. All adult patients admitted to the Mayo Clinic in Rochester, Minnesota, between January 2009 and December 2019 who had community-acquired pneumonia (CAP) underwent screening. The presence of CAP was determined using International Classification of Diseases 9 (481–486) and 10 (J13, J15, and J18) codes, coupled with note searches. Community-acquired pneumonia was defined as an acute infection of the lung parenchyma exhibiting clinical symptoms (cough, fever, pleuritic chest pain, and dyspnea) and a new radiographic infiltrate not being acquired in the hospital or healthcare setting [14].

The exclusion criteria were:

- The absence of Minnesota research authorization;
- Baseline conditions of human immune deficiency virus infection, interstitial lung disease, leukopenia, or neutropenia;
- Diagnoses of hospital-acquired pneumonia, ventilator-associated pneumonia, or aspiration pneumonia;
- Readmissions (only the earliest admission during the study period was included per individual);
- Hospital stays under 24 h,
- The absence of neutrophil or lymphocyte levels within the initial 24 h of admission (Figure 1).

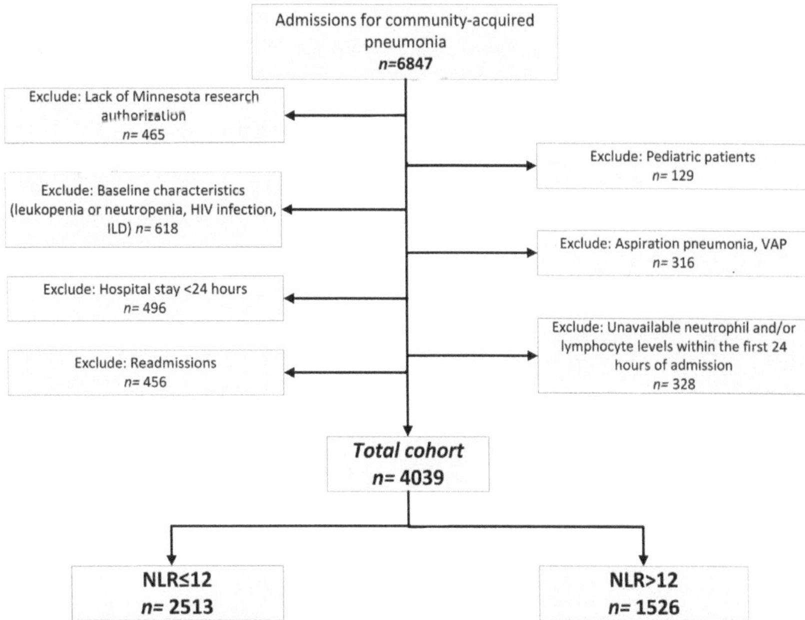

Figure 1. Flowchart of the identification of the patients in neutrophil/lymphocyte ratio analyses. HIV: human immunodeficiency virus, ILD: interstitial lung disease, NLR: neutrophil/lymphocyte ratio, VAP: ventilator-associated pneumonia.

2.2. Variable Definitions

The data were extracted by the Anesthesia Clinical Research Unit team using standardized and validated queries [15–17]. Extracted data included relevant demographic information, comorbidities (recorded as the presence of individual comorbidities alongside the Charlson Comorbidity Index [CCI]) [18], and disease severity indexes, evaluated via clinical scores. The medication exposure evaluation was limited to corticosteroids and vasopressors.

The white blood cell count, which measures the total number of leukocytes, is recorded as 10^9/L in our institution, with the institutional normal falling between 3.4 and 9.6×10^9/L. As per the differential components, the normal ranges for neutrophil and lymphocyte counts are considered to be 1.56 to 6.45×10^9/L and 0.95 to 3.07×10^9/L, respectively. Neutrophil and lymphocyte levels tested during the first 24 h of admission (the first one available if more than one result was available during the timeline) were recorded as continuous variables and used to calculate the neutrophil/lymphocyte ratio (NLR). Based on the literature, an a priori NLR cut-off of 12 was chosen (patients were compared as those whose NLR level was $\leq$12 vs. those >12) [13].

2.3. Outcomes

Primary outcomes included mortality during hospitalization and within six months of admission. Secondary outcomes encompassed the need for invasive and non-invasive mechanical ventilation (IMV) (NIMV), intensive care unit (ICU) admission, and hospital- and ICU-free days, calculated as days alive spent outside of the hospital or the ICU, respectively, within 28 days of admission, resulting in 0 for patients who died during the stay or had a length of stay of $\geq$28 days [19]. In a similar fashion, IMV-free days were also calculated and evaluated. The prediction of in-hospital mortality and the requirement for ICU admission using PSI-only and PSI combined with NLR were also calculated.

2.4. Statistical Analysis

The median, interquartile range (IQR) for continuous data and the numbers and frequencies for categorical variables were used to perform and present descriptive summary statistics. The chi-square test was used to examine categorical data, while the Mann–Whitney U test was used to evaluate continuous variables. The NLR ≤ 12 group was used as the reference for comparing the two groups. Results were reported using odds ratios (ORs), p values, estimates, and 95% confidence intervals (CIs). The analyses were adjusted for the baseline conditions and severity of the patients, measured by the PSI score using binary logistic regression and linear regression, depending on the characteristics of the outcomes of interest. As PSI already includes baseline characteristics such as age, gender, and certain comorbid conditions, no additional covariates were deemed necessary for the multivariable calculations comparing the NLR groups.

To better understand NLR's association with outcomes, the following sensitivity analyses were conducted:

- Evaluating the impact of increasing NLR as a continuous variable;
- Classifying patients into four categories based on quartiles (quartile [Q] #1 NLR < 5.13, Q #2 NLR $\geq$ 5.13 and <9.26, Q #3 NLR $\geq$ 9.26 and <16.35, Q #4 NLR $\geq$ 16.35; Q #1 was the reference group);
- Limiting the analyses to patients with neutrophilia (defined as having an absolute neutrophil count of $>6.45 \times 10^9$, the institutional cut-off level);
- Limiting the analyses to patients with lymphopenia (defined as having an absolute lymphocyte count of $<0.95 \times 10^9$, the institutional cut-off level).

Finally, C-statistics was used to examine the NLR and PSI's predictive ability for mortality and ICU admission. This included creating a receiver operating characteristic (ROC) curve and determining the area under the curve (AUC), as well as the 95% CI The potential contribution of NLR to the PSI was also evaluated. The ROC curves for both approaches were compared using DeLong's test [20]. A two-sided p-value of 0.05 was considered significant. IBM SPSS v27.0 (IBM Statistical Package for Social Sciences Statistics for Windows, Armonk, NY, USA) and MedCalc Statistical Software v19.1 (MedCalc Software bv, Ostend, Belgium) were used to perform the calculations.

3. Results

Following the application of exclusion criteria, 4039 patients out of the initial 6847 were included in the cohort (Figure 1).

3.1. Comparison of Patients with an NLR $\leq$ 12 vs. >12

The primary analysis compared 2513 patients with NLR $\leq$ 12 and 1526 with NLR > 12. Table 1 illustrates the distribution of baseline characteristics across both groups. Although there were some differences in the distribution of specific comorbidities, the baseline comorbidity burdens were generally well-balanced, as indicated by a comparable CCI (median [IQR] = 7 [5,10] and 7 [5,9] for NLR $\leq$ 12 and >12 groups, respectively, $p = 0.541$).

Table 2 outlines the distribution of the primary outcomes and the results of the multivariable analysis. The odds of ICU admission were higher in patients with NLR > 12 ($n = 710$, 46.5%) compared to those with NLR $\leq$ 12 ($n = 869$, 34.6%) (adjusted OR [95% CI] = 1.41 [1.22, 1.62]).

Table 1. Demographics and clinical characteristics of patients.

Variables	Total (n = 4039)	NLR ≤ 12 (n = 2513)	NLR > 12 (n = 1526)	p, OR (95% CI)
Age, median (IQR)	78 (65, 88)	77 (65, 88)	80 (67, 88)	0.003
Sex, Male, no. (%)	2172 (53.8%)	1293 (51.5%)	879 (57.6%)	<0.001, 1.28 (1.13, 1.46)
Race, no. (%)				0.038
White	3805 (94.2%)	2349 (93.5%)	1456 (95.4%)	
Asian	42 (1.0%)	31 (1.2%)	11 (0.7%)	
African American	64 (1.6%)	50 (2.0%)	14 (0.9%)	
Others	128 (3.2%)	83 (3.3%)	45 (2.9%)	
Ethnicity, no. (%)				0.401
Non-Hispanic	3897 (96.5%)	2423 (96.4%)	1474 (96.6%)	
Hispanic	65 (1.6%)	45 (1.8%)	20 (1.3%)	
Unknown/Other	77 (1.9%)	45 (1.8%)	32 (2.1%)	
Comorbidities, no. (%)				
Congestive heart failure	1118 (27.7%)	703 (28.0%)	415 (27.2%)	0.592, 0.96 (0.83, 1.11)
COPD	1310 (32.4%)	740 (29.4%)	570 (37.4%)	<0.001, 1.43 (1.25, 1.63)
Diabetes	1220 (30.2%)	790 (31.4%)	430 (28.2%)	0.029, 0.86 (0.74, 0.98)
Chronic kidney disease	1159 (28.7%)	724 (28.8%)	435 (28.5%)	0.836, 0.99 (0.86, 1.13)
Malignancy	938 (23.5%)	576 (22.9%)	372 (24.4%)	0.290, 1.08 (0.93, 1.26)
Neutrophil/lymphocyte ratio, median, (IQR)	9.26 (5.13, 16.35)	6.09 (3.80, 8.67)	19.71 (14.86, 27.88)	
Clinical Severity Scores				
Pneumonia Severity Index, median (IQR)	113 (87, 141)	109 (83, 136)	119 (93, 148)	<0.001
CURB 65	3 (2, 3)	3 (2, 3)	3 (2, 4)	<0.001
APACHE III	65 (52, 80)	64 (51, 79)	66 (53, 81)	0.013
Need for vasopressor during hospitalization, no. (%)	888 (22.0%)	512 (20.4%)	376 (24.6%)	0.002, 1.28 (1.10, 1.49)

APACHE: Acute Physiology and Chronic Health Evaluation, CI: confidence interval, COPD: Chronic Obstructive Pulmonary Disease NLR: neutrophil/lymphocyte ratio, IQR: interquartile range, OR: odds ratio.

Table 2. Primary and secondary outcomes based on NLR levels.

Variables	Total (n = 4039)	NLR ≤ 12 (n = 2513)	NLR > 12 (n = 1526)	Univariate Analysis		Multivariate Analysis *	
				Odds Ratio (95% CI)	p Value	Odds Ratio (95% CI)	p Value
Hospital mortality, no. (%)	128 (3.2%)	69 (2.7%)	59 (3.9%)	1.43 (1.00, 2.03)	0.049	1.12 (0.77, 1.61)	0.559
Mortality at 6 months, no. (%)	638 (15.8%)	375 (14.9%)	263 (%17.2)	1.19 (1.00, 1.41)	0.051	0.99 (0.83, 1.19)	0.923
Need for invasive mechanical ventilation, no. (%)	704 (17.4%)	406 (16.2%)	298 (19.5%)	1.26 (1.07, 1.49)	0.006	1.08 (0.91,1.28)	0.395
Need for invasive and non-invasive mechanical ventilation, no. (%)	1348 (33.4%)	789 (31.4%)	559 (36.6%)	1.26 (1.11, 1.44)	<0.001	1.08 (0.93,1.24)	0.312
ICU admission	1579 (39.1%)	869 (34.6%)	710 (46.5%)	1.65 (1.45, 1.87)	<0.001	1.41 (1.22, 1.62)	<0.001
					p value	Estimate (95% CI)	p value
Hospital-free days, median (IQR)	24.01 (21.04, 25.44)	24.13 (21.38, 25.65)	23.78 (20.23, 25.29)		<0.001	−0.34 (−0.72, 0.04)	0.081
Invasive ventilator-free days, median (IQR)	26.29 (23.67, 27.27)	26.19 (23.19, 27.28)	26.40 (23.97, 27.19)		0.510	0.96 (−0.31, 2.23)	0.139

CI: confidence interval, ICU: intensive care unit, IQR: interquartile range, NLR: neutrophil/lymphocyte ratio. * Data were analyzed using multivariable regression models adjusting for pneumonia severity index. NLR ≤ 12 was the reference.

3.2. *Sensitivity Analyses*

3.2.1. Evaluation of the Impact of the NLR as a Continuous Variable

Upon analyzing the relationship between an increasing NLR and various outcomes without categorization, we observed a rise in the odds of both hospital mortality and ICU admission as the NLR increased (adjusted OR [95% CI] = 1.01 [1.00, 1.02], $p = 0.047$ and adjusted OR [95% CI] = 1.02 [1.01, 1.02], $p < 0.001$ for hospital mortality and ICU admission, respectively). Additionally, higher NLR levels were associated with a significant decrease in the number of hospital-free days (adjusted estimate [95% CI] = −0.02 [−0.04, −0.01]). Although univariate analyses suggested a significant association between the NLR and mortality within six months after admission, as well as the need for mechanical ventilation, these associations did not remain significant after adjusting for the PSI (adjusted OR [95% CI] = 1.0 [1.0, 1.01], $p = 0.693$ and adjusted OR [95% CI] = 1.0 [1.0, 1.01], $p = 0.198$ for mortality at six months and the requirement for mechanical ventilation, respectively).

3.2.2. Four-Group Comparison Based on NLR Quartiles

Supplementary Table S1 and Table 3 illustrate the distribution of baseline characteristics and outcomes among the four groups of patients based on NLR levels. Similar to the primary analyses, the need for ICU admission was significantly different in patients belonging to Q#3 ($n = 405$, 40.1%) and Q#4 ($n = 492$ (48.7%) compared to Q#1 ($n = 319$, 31.6%) (adjusted OR [95% CI] = 1.38 [1.16, 1.69] and 1.64 [1.34, 2.01] for Q#3 and Q#4, respectively).

Table 3. Primary and secondary outcomes of patients according to their neutrophil/lymphocyte ratio levels *.

Variables	Quartile #1 (n = 1010)	Quartile #2 (n = 1010)	Quartile #3 (n = 1009)	Quartile #4 (n = 1010)	Univariable Analysis	Multivariable Analysis **	
					p Value	Odds Ratio (95% CI)	p Value
Hospital mortality, no. (%)	24 (2.4)	27 (2.7)	34 (3.4)	43 (4.3)	0.075	Q#2 1.09 (0.62, 1.93)	0.763
						Q#3 1.3 (0.76, 2.23)	0.343
						Q#4 1.28 (0.76, 2.16)	0.352
Mortality at 6 months, no. (%)	162 (16)	144 (14.3)	152 (15.1)	180 (17.8)	0.147	Q#2 0.33 (0.64, 1.07)	0.149
						Q#3 0.84 (0.66, 1.08)	0.185
						Q#4 0.86 (0.68, 1.1)	0.238
Need for invasive mechanical ventilation, no. (%)	152 (15)	165 (16.3)	185 (18.3)	202 (20)	0.018	Q#2 1.07 (0.84, 1.37)	0.581
						Q#3 1.19 (0.93, 1.51)	0.164
						Q#4 1.12 (0.88, 1.43)	0.341
Need for invasive and non-invasive mechanical ventilation, no. (%)	300 (29.7)	324 (32.1)	346 (34.3)	378 (37.4)	0.002	Q#2 1.08 (0.89, 1.32)	0.422
						Q#3 1.15 (0.95, 1.4)	0.158
						Q#4 1.12 (0.92, 1.37)	0.254
ICU admission	319 (31.6)	363 (35.9)	405 (40.1)	492 (48.7)	<0.001	Q#2 1.19 (0.97, 1.46)	0.090
						Q#3 1.38 (1.13, 1.69)	0.002
						Q#4 1.64 (1.34, 2.01)	<0.001
					p value	Estimate (95% CI)	p value
Hospital-free days, median (IQR)	24.3 (22.06, 25.88)	24.01 (21.12, 25.48)	23.98 (21, 25.41)	23.55 (19.96, 25.23)	<0.001	Q#2 −0.41 (−0.93, 0.11)	0.120
						Q#3 −0.43 (−0.95, 0.09)	0.104
						Q#4 −0.63 (−1.15, −0.11)	0.017

CI: confidence interval, ICU: intensive care unit, IQR: interquartile range, Q: quartile. * Neutrophil-lymphocyte ratio levels per quartiles: Quartile #1, <5.13; Quartile #2, ≥5.13 and <9.26; Quartile #3, ≥9.26 and <16.35; Quartile #4, ≥16.35. ** Data were analyzed using multivariable regression models adjusting for pneumonia severity index. Quartile #1 was the reference.

3.2.3. Analysis among Patients with Neutrophilia

Supplementary Tables S2 and S3 depict the distribution of baseline characteristics and outcomes for the comparative analysis among the 2989 patients with neutrophilia.

3.2.4. Analysis among Patients with Lymphopenia

Supplementary Tables S4 and S5 showcase the distribution of baseline characteristics and outcomes for the comparative analysis among the 1905 patients with lymphopenia.

3.3. Discriminatory Performance

Supplementary Table S6 presents the distribution of baseline characteristics between patients who died during their hospital stay and those who were discharged alive. The distinguishing capacity of the NLR for patients who died during hospitalization was poor (AUC [95% CI] = 0.57 [0.52, 0.62]). The discriminatory capacity of the PSI for patients who died during hospitalization was acceptable (AUC [95% CI] = 0.78 [0.75, 0.82]) and did not considerably improve with the addition of the NLR (AUC [95% CI] = 0.78 [0.75, 0.82]) (Figure 2a), with a negligible difference between the AUC for PSI versus PSI and NLR of 0.0025 (95% CI: −0.0039, 0.0089, p = 448).

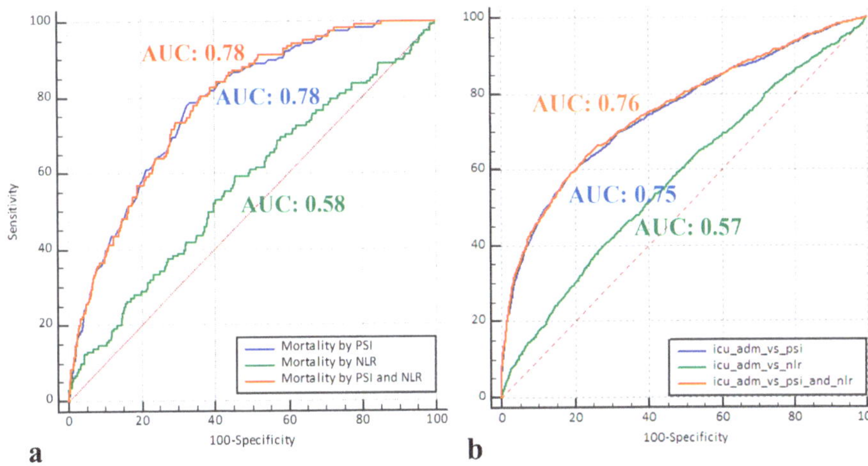

Figure 2. Receiver operating characteristic curves for pneumonia severity index (PSI) only, neutrophil/lymphocyte ratio (NLR) only, and PSI combined with NLR in detecting (**a**) in-hospital mortality (the area under the receiver operating characteristic curve (AUC) was 0.78 (95% CI, 0.75–0.82) for PSI only and PSI combined with NLR approaches, while it was 0.57 (95% CI, 0.52–0.62) for NLR only) and (**b**) the requirement for intensive care unit admission (the AUC was 0.75 (95% CI, 0.74–0.77), 0.58 (95% CI, 0.56–0.6), and 0.76 (95% CI: 0.74–0.77) for PSI only, NLR only, and PSI combined with NLR, respectively). ICU: intensive care unit.

The discriminatory capacity of the NLR for patients requiring an ICU admission was poor (AUC [95% CI] = 0.58 [0.56, 0.6]). The distinguishing capacity of the PSI for patients requiring an ICU stay was acceptable (AUC [95% CI] = 0.75 [0.74, 0.77]) and significantly improved with the addition of the NLR (AUC [95% CI] = 0.76 [0.74, 0.77]) (Figure 2b), with a difference between the AUC for PSI versus PSI and NLR of 0.0041 (95% CI: 0.001, 0.007, p = 0.008).

4. Discussion

In this large single-center retrospective cohort of hospitalized patients with community-acquired pneumonia, we found that a higher NLR (>12) was not associated with an increased odds for in-hospital mortality and 6-month mortality. Notably, there was a significant difference in the severity of illness as measured by the PSI, CURB-65, and APACHE III scores between the subgroups (NLR $\leq$ 12 and >12), and a significant association with mortality following univariate analysis. However, after adjusting for baseline comorbid conditions and the severity of illness using the PSI, the association between a higher NLR and mortality became insignificant. Additional adjusted analysis with the NLR as a continuous variable revealed a barely significant association between the NLR and in-hospital mortality. On assessment of secondary outcomes, a higher NLR was only found to be associated with an increased odds of ICU admission in this cohort, while no association was observed with other secondary outcomes including the need for invasive and non-invasive mechanical ventilation. Further adjusted analysis with the NLR as a continuous variable also revealed a significant association with ICU admission and hospital-free days, but no association with other secondary outcomes. These findings were consistent on additional sensitivity analysis evaluating outcomes in NLR quartiles, with an increasing odds of ICU admission in the third and fourth quartiles. Interestingly, the increased odds of ICU admission were only observed in the subset of patients with a high NLR and co-existing lymphopenia compared to patients with concomitant neutrophilia. Additionally, an increased odds for a need of both invasive and non-invasive mechanical ventilation were noted in this lymphopenic subgroup. Potential explanations include the presence of other confounding factors like the increasing use of non-invasive mechanical support for high-risk extubation patients such as COPD. Finally, in this cohort, the discriminatory capacity of the NLR in predicting both mortality and ICU admission was inferior to the PSI. Moreover, the addition of the NLR to the PSI did not improve the prediction of the mortality model. The improvement with the addition of the NLR to the PSI for predicting ICU admission, albeit statistically significant, was very small and not clinically meaningful.

The utility of the NLR in predicting adverse outcomes in CAP has been previously reported in other studies [7,8,10,11,21,22]. However, these studies reported conflicting results and were limited by sample size. Further, only a few studies confirmed an association between the NLR and mortality by multivariate analysis. In a recent systematic review of nine studies (n = 3340), the association between a higher NLR and mortality was observed to be significant. An NLR cut-off value >10 in this systematic review was found to predict mortality compared to the PSI, CURB-65, and other biomarkers including C-reactive protein, lymphocytes, neutrophils, and WBC [13]. A higher sensitivity and specificity were observed at a cut-off value between 11.2 and 13.4. In contrast, our results demonstrated a barely significant association between an elevated NLR and mortality, and also a poor discriminatory ability of the NLR in predicting mortality compared to the PSI. Although the results of a retrospective study by Postma et al., the largest study included in the systematic review (n = 1549), reported a significant association between the NLR and mortality based on bivariate analysis, this was absent in the subsequent multivariate analysis. In addition, the lack of improvement in the PSI model with the addition of the NLR, as seen in our study, was similarly observed in the large study by Postma et al. [12]. Therefore, our result in this larger cohort further limits the clinical applicability of utilizing the NLR as a prognostic tool for mortality, alone or in combination with severity scores, in CAP. Further studies, including prospective studies and an updated systematic review will be needed.

In comparison with other prior studies evaluating the prognostic value of the NLR in CAP, our study further found a significant association between the NLR and need for ICU admission, an important outcome that is rarely assessed. In addition, other outcomes, including the need for mechanical ventilation, hospital-free days, and ventilator-free days, were all assessed. Moreover, our study explored the outcomes in subset of patients with neutrophilia versus lymphopenia, with significant associations observed in the lymphopenia group.

The strengths of our study compared to prior similar studies include the large sample size (the largest sample size to our knowledge), the multivariable analysis adjusting for the severity of illness and comorbidities, the exploration of other clinical outcomes in addition to mortality, and the sensitivity analysis evaluating specific subgroups. As this was a single-center cohort study, findings reported are reflective of characteristics observed in a large academic center. Other limitations include the potential bias existing within the dataset, missing data, the lack of microbiological data, and other unmeasured confounders.

5. Conclusions

In this study, the neutrophil/lymphocyte ratio was not superior to the pneumonia severity index for predicting in-hospital mortality in patients who were hospitalized with community-acquired pneumonia. This limits its applicability as a prognostic enrichment tool, but additional studies are needed for assessing its use as a predictive enrichment tool.

Supplementary Materials: The following supporting information can be downloaded at https://www.mdpi.com/article/10.3390/biomedicines12020260/s1: Table S1: Demographics and clinical characteristics of patients according to their neutrophil/lymphocyte ratio levels; Table S2: Demographics and clinical characteristics of patients who had neutrophilia; Table S3: Primary and secondary outcomes of patients who had neutrophilia based on NLR; Table S4: Demographics and clinical characteristics of patients who had lymphopenia; Table S5: Primary and secondary outcomes of patients who had lymphopenia based on NLR; Table S6: Demographics and clinical characteristics of patients according to their in-hospital mortality status.

Author Contributions: Conceptualization, A.T., F.W.W., O.G. and Y.E.O.; methodology, A.T. and Y.E.O.; software, A.T.; validation, A.T. and Y.E.O.; formal analysis, A.T.; investigation, A.T. and Y.E.O.; resources, A.T., O.G. and Y.E.O.; data curation, Y.E.O.; writing—original draft preparation, A.T. and F.W.W.; writing—review and editing, O.G. and Y.E.O.; visualization, A.T.; supervision, O.G. and Y.E.O.; funding acquisition, Y.E.O. All authors have read and agreed to the published version of the manuscript.

Funding: This research was supported by the National Institute of Health, grant number K23HL168212; the Robert D. and Patricia E. Kern Center for the Science of Health Care Delivery, Mayo Clinic; and the Robert A. Winn Diversity in Clinical Trials Career Development Award.

Institutional Review Board Statement: The institutional review board (IRB) at the Mayo Clinic approved this research as low-risk, denoted by the IRB number 17-011140, approval date: 20 December 2017, with a waiver of the requirement for written informed consent.

Informed Consent Statement: Patient consent was waived due to the study being conducted on deidentified retrospective data.

Data Availability Statement: The data that support the findings of this study are available from the corresponding author, A.T., upon reasonable request.

Conflicts of Interest: The authors declare no conflicts of interest.

References

1. Hayes, B.H.; Haberling, D.L.; Kennedy, J.L.; Varma, J.K.; Fry, A.M.; Vora, N.M. Burden of pneumonia-associated hospitalizations: United States, 2001–2014. *Chest* **2018**, *153*, 427–437. [CrossRef] [PubMed]
2. Ramírez, P.; Ferrer, M.; Martí, V.; Reyes, S.; Martínez, R.; Menéndez, R.; Ewig, S.; Torres, A. Inflammatory biomarkers and prediction for intensive care unit admission in severe community-acquired pneumonia. *Crit. Care Med.* **2011**, *39*, 2211–2217. [CrossRef] [PubMed]
3. Moons, K.G.; Altman, D.G.; Vergouwe, Y.; Royston, P. Prognosis and prognostic research: Application and impact of prognostic models in clinical practice. *BMJ* **2009**, *338*, b606. [CrossRef] [PubMed]
4. Aujesky, D.; Fine, M.J. The pneumonia severity index: A decade after the initial derivation and validation. *Clin. Infect. Dis. Off. Publ. Infect. Dis. Soc. Am.* **2008**, *47* (Suppl. S3), S133–S139. [CrossRef] [PubMed]
5. Mandell, L.A.; Wunderink, R.G.; Anzueto, A.; Bartlett, J.G.; Campbell, G.D.; Dean, N.C.; Dowell, S.F.; File, T.M., Jr.; Musher, D.M.; Niederman, M.S.; et al. Infectious Diseases Society of America/American Thoracic Society consensus guidelines on the management of community-acquired pneumonia in adults. *Clin. Infect. Dis. Off. Publ. Infect. Dis. Soc. Am.* **2007**, *44* (Suppl. S2), S27–S72. [CrossRef] [PubMed]
6. Ewig, S.; Ruiz, M.; Mensa, J.; Marcos, M.A.; Martinez, J.A.; Arancibia, F.; Niederman, M.S.; Torres, A. Severe community-acquired pneumonia. Assessment of severity criteria. *Am. J. Respir. Crit. Care Med.* **1998**, *158*, 1102–1108. [CrossRef] [PubMed]

7. Ozmen, I.; Karakurt, Z.; Salturk, C.; Kargin, F.; Takir, H.B.; Aksoy, E.; Sari, R.; Celik, E.; Tuncay, E.A.; Yildirim, E.; et al. Can N-terminal pro B-type natriuretic peptide, neutrophil-to-lymphocyte ratio, C-reactive protein help to predict short and long term mortality? *Bratisl. Lek. Listy.* **2016**, *117*, 587–594. [CrossRef] [PubMed]
8. Yang, T.; Wan, C.; Wang, H.; Qin, J.; Chen, L.; Shen, Y.; Wen, F. The prognostic and risk-stratified value of neutrophil–lymphocyte count ratio in Chinese patients with community-acquired pneumonia. *Eur. J. Inflamm.* **2017**, *15*, 22–27. [CrossRef]
9. Kaya, Y.; Taş, N.; Çanakçı, E.; Cebeci, Z.; Özbilen, M.; Keskin, H.; Yıldırım, B.B. Relationship of neutrophil-to-lymphocyte ratio with presence and severity of pneumonia. *J. Clin. Anal. Med.* **2018**, *9*, 452–457.
10. de Jager, C.P.; Wever, P.C.; Gemen, E.F.; Kusters, R.; van Gageldonk-Lafeber, A.B.; van der Poll, T.; Laheij, R.J. The neutrophil-lymphocyte count ratio in patients with community-acquired pneumonia. *PLoS ONE* **2012**, *7*, e46561. [CrossRef] [PubMed]
11. Cataudella, E.; Giraffa, C.M.; Di Marca, S.; Pulvirenti, A.; Alaimo, S.; Pisano, M.; Terranova, V.; Corriere, T.; Ronsisvalle, M.L.; Di Quattro, R.; et al. Neutrophil-To-Lymphocyte Ratio: An Emerging Marker Predicting Prognosis in Elderly Adults with Community-Acquired Pneumonia. *J. Am. Geriatr. Soc.* **2017**, *65*, 1796–1801. [CrossRef] [PubMed]
12. Postma, D.F.; van Werkhoven, C.H.; Schuurman, J.; van Elden, L.J.; Oosterheert, J.J.; Bonten, M.J. Prognostic value of the neutrophil/lymphocyte ratio in patients hospitalized with community-acquired pneumonia. *Adults* **2016**, 155.
13. Kuikel, S.; Pathak, N.; Poudel, S.; Thapa, S.; Bhattarai, S.L.; Chaudhary, G.; Pandey, K.R. Neutrophil-lymphocyte ratio as a predictor of adverse outcome in patients with community-acquired pneumonia: A systematic review. *Health Sci. Rep.* **2022**, *5*, e630. [CrossRef] [PubMed]
14. Odeyemi, Y.E.; Lal, A.; Barreto, E.F.; LeMahieu, A.M.; Yadav, H.; Gajic, O.; Schulte, P. Early machine learning prediction of hospitalized patients at low risk of respiratory deterioration or mortality in community-acquired pneumonia: Derivation and validation of a multivariable model. *Biomol. Biomed.* **2023**. [CrossRef] [PubMed]
15. Herasevich, V.; Kor, D.J.; Li, M.; Pickering, B.W. ICU data mart: A non-iT approach. A team of clinicians, researchers and informatics personnel at the Mayo Clinic have taken a homegrown approach to building an ICU data mart. *Health. Inf.* **2011**, *28*, 42–45.
16. Clinic, M. *Mayo Data Explorer*; 2021.
17. Clinic, M. *Advanced Text Explorer*; 2021.
18. Charlson, M.E.; Pompei, P.; Ales, K.L.; MacKenzie, C.R. A new method of classifying prognostic comorbidity in longitudinal studies: Development and validation. *J. Chronic. Dis.* **1987**, *40*, 373–383. [CrossRef] [PubMed]
19. Schoenfeld, D.A.; Bernard, G.R. Statistical evaluation of ventilator-free days as an efficacy measure in clinical trials of treatments for acute respiratory distress syndrome. *Crit. Care Med.* **2002**, *30*, 1772–1777. [CrossRef] [PubMed]
20. DeLong, E.R.; DeLong, D.M.; Clarke-Pearson, D.L. Comparing the areas under two or more correlated receiver operating characteristic curves: A nonparametric approach. *Biometrics* **1988**, *44*, 837–845. [CrossRef] [PubMed]
21. Avci, S.; Perincek, G. The alveolar-arterial gradient, pneumonia severity scores and inflammatory markers to predict 30-day mortality in pneumonia. *Am. J. Emerg. Med.* **2020**, *38*, 1796–1801. [CrossRef]
22. Curbelo, J.; Luquero Bueno, S.; Galván-Román, J.M.; Ortega-Gómez, M.; Rajas, O.; Fernández-Jiménez, G.; Vega-Piris, L.; Rodríguez-Salvanes, F.; Arnalich, B.; Díaz, A.; et al. Inflammation biomarkers in blood as mortality predictors in community-acquired pneumonia admitted patients: Importance of comparison with neutrophil count percentage or neutrophil-lymphocyte ratio. *PLoS ONE* **2017**, *12*, e0173947. [CrossRef] [PubMed]

biomedicines

MDPI

Review

Recent Data about the Use of Corticosteroids in Sepsis—Review of Recent Literature

Alexandra Lazar

Anesthesiology and Intensive Care Department, "George Emil Palade" University of Medicine, Pharmacy, Science and Technology from Tirgu Mures, 540142 Targu Mures, Romania; alexandra.lazar@umfst.ro

Abstract: Sepsis, characterized by life-threatening organ dysfunction due to a maladaptive host response to infection, and its more severe form, septic shock, pose significant global health challenges. The incidence of these conditions is increasing, highlighting the need for effective treatment strategies. This review explores the complex pathophysiology of sepsis, emphasizing the role of the endothelium and the therapeutic potential of corticosteroids. The endothelial glycocalyx, critical in maintaining vascular integrity, is compromised in sepsis, leading to increased vascular permeability and organ dysfunction. Corticosteroids have been used for over fifty years to treat severe infections, despite ongoing debate about their efficacy. Their immunosuppressive effects and the risk of exacerbating infections are significant concerns. The rationale for corticosteroid use in sepsis is based on their ability to modulate the immune response, promote cardiovascular stability, and potentially facilitate organ restoration. However, the evidence is mixed, with some studies suggesting benefits in terms of microcirculation and shock reversal, while others report no significant impact on mortality or organ dysfunction. The Surviving Sepsis Campaign provides cautious recommendations for their use. Emerging research highlights the importance of genomic and transcriptomic analyses in identifying patient subgroups that may benefit from corticosteroid therapy, suggesting a move toward personalized medicine in sepsis management. Despite potential benefits, the use of corticosteroids in sepsis requires careful consideration of individual patient risk profiles, and further research is needed to optimize their use and integrate genomic insights into clinical practice. This review underscores the complexity of sepsis treatment and the ongoing need for evidence-based approaches to improve patient outcomes.

Keywords: corticosteroids; sepsis; septic shock; critically ill; intensive care

Citation: Lazar, A. Recent Data about the Use of Corticosteroids in Sepsis—Review of Recent Literature. *Biomedicines* **2024**, *12*, 984. https://doi.org/10.3390/biomedicines12050984

Academic Editors: Wen-Lin Su, Sheng-Kang Chiu, Chih-Hao Shen and Yi-Ting Chen

Received: 28 March 2024
Revised: 23 April 2024
Accepted: 28 April 2024
Published: 30 April 2024

1. Introduction

Sepsis is recognized as a critical condition characterized by life-threatening organ dysfunction resulting from a maladaptive host response to infection. Septic shock is a more severe form of sepsis, distinguished by profound circulatory, cellular, and metabolic dysfunctions that significantly elevate mortality risk beyond that of sepsis alone [1]. These conditions are part of a growing global health crisis, presenting complex challenges for emergency physicians due to their rising incidence and the intricate interplay of pathophysiological, molecular, genetic, and clinical factors [2]. Since the initial consensus definition in 1991 (Sepsis-1), the global occurrence of sepsis and septic shock has been on an upward trajectory, with estimates indicating around 49 million sepsis cases and 11 million sepsis-related fatalities worldwide in 2017 [3].

Currently, the development of sepsis is understood to involve a complex interplay of numerous pro-inflammatory and anti-inflammatory mediators. Recent advances have also shed light on cellular changes specific to sepsis, with a growing emphasis on the role of microcirculation in the transition from sepsis to septic shock. In this framework, the endothelium emerges as a pivotal element in sepsis pathophysiology due to its critical functions in controlling microcirculation and regulating coagulation, as well as inflammatory and anti-inflammatory pathways [4,5]. The endothelial glycocalyx, made up of proteoglycans and glycoproteins, plays multiple roles, including acting as a physical barrier to control vascular permeability, facilitating leukocyte and platelet adhesion, and moderating inflammatory and anti-inflammatory responses. However, the structural integrity of the glycocalyx can be compromised—referred to as "glycocalyx shedding"—through exposure to oxidants, cytokines, exotoxins, and endotoxins from bacteria. This degradation leads to the migration of leukocytes across the endothelium and heightened vascular permeability, resulting in edema that increases interstitial pressure and exacerbates the impairment of tissue perfusion [6].

2. Corticosteroids as a Therapeutic Option

Over fifty years since the inaugural randomized, controlled trial examining use of corticosteroids for severe infections, their widespread use among physicians continues despite significant disagreements among experts regarding their risk-to-benefit ratio [7]. The widespread and enduring use of corticosteroids in treating severe infections can be attributed to the immediate and observable improvement in critical conditions like shock and respiratory failure observed in clinical settings. The ongoing debate among experts over the use of corticosteroids is not driven by new scientific evidence but by differing interpretations of existing data.

Corticosteroids are recognized for their extensive immunosuppressive properties, which can elevate the risk of infectious complications for patients. Moreover, stress often triggers an immediate surge in inflammatory mediators, leading to a systemic inflammatory response and significant suppression of the immune system [8].

The Rationale for Using Corticosteroids

- Overactivity of proinflammatory pathways relative to endogenous glucocorticoid activity

Experts widely agree that uncontrolled systemic inflammation is a fundamental aspect of severe sepsis, significantly influencing the advancement of organ failure and death. The management of inflammation involves a complex interaction between the neuroendocrine and immune systems [9]. On a microscopic scale, equilibrium within the body is maintained through the interplay of two dynamic systems. The nuclear factor kappa B (NF-κB) pathway triggers the release of substances that promote inflammation, while the synergy between glucocorticoids and the glucocorticoid receptor alpha (G-GRα) complex serves to suppress inflammatory reactions [10]. In all cells, these processes are typically dormant; the NF-κB pathway is restrained by its inhibitor, inhibitory factor kappa B (I-κB), and regulation of the G-GRα complex is achieved by maintaining an equilibrium between its alpha and beta forms. When these systems successfully counteract each other, stability is preserved. Nonetheless, an inclination toward the activation of NF-κB may result in uncontrolled inflammation [10,11].

In cases of extended acute respiratory distress syndrome or septic shock in patients, a disproportion has been observed, with NF-κB being excessively active about the G-GRα complex. This imbalance is recognized as a contributing element to the damage of cells, tissues, and organs. The root reasons for adrenal insufficiency during critical illnesses have been extensively investigated in separate research [12]. Influences including drugs that alter cortisol metabolism, harm to the hypothalamic–pituitary–adrenal axis, and stimulation of the inducible nitric oxide synthase (iNOS) enzyme through cytokines contribute to the death of neuroendocrine cells. This sequence of events subsequently leads to a decreased responsiveness to ACTH (adrenocorticotropic hormone) and glucocorticoids. It has been noted that people who overcome critical illnesses typically regain normal endocrine function within a few weeks to months following their hospital discharge [12].

- Molecular mechanisms through which glucocorticoids act align seamlessly with the underlying pathological processes of sepsis

Glucocorticoids function via both genomic and non-genomic pathways. The non-genomic effects, manifesting within minutes after exposure to glucocorticoids, include decreased platelet aggregation and cell adhesion, reduced activity of intracellular phosphotyrosine kinases, and increased expression of annexin 1 on the cell surface [13]. The rapid effects are believed to result from glucocorticoids interacting with locations on the cell membrane. Conversely, the genomic activities of glucocorticoids are generally divided into two categories: transrepression and transactivation effects [14]. The prevailing view is that glucocorticoids primarily modulate rather than simply suppress immune cell function [15].

Microarray studies have shown that exposure to glucocorticoids activates a greater number of genes than it represses, highlighting their intricate impact on gene expression. Further research has also demonstrated that glucocorticoids encourage the development of specific anti-inflammatory monocyte subtypes, which rapidly migrate to areas of inflammation. Moreover, glucocorticoids extend the survival of these monocytes by activating anti-apoptotic pathways through the A3 adenosine receptor. This detailed function of glucocorticoids in regulating immune responses proves particularly advantageous in controlling the excessive inflammation seen in sepsis [16,17].

- Cardiovascular stability in sepsis

Corticosteroids facilitate the retention of sodium by affecting both mineralocorticoid and glucocorticoid receptors, a process that helps to combat the hypovolemia observed in the early phases of sepsis. Moreover, corticosteroids contribute to the retention of sodium and water within the blood vessel walls, which increases systemic vascular resistance. Through non-genomic actions, corticosteroids can quickly boost the responsiveness of blood vessels to alpha agonists in a matter of minutes to hours, leading to elevated mean arterial pressure and systemic vascular resistance [18]. Corticosteroids also contribute to blocking ATP-dependent potassium channels. Their ability to dampen the gene expression for iNOS and cyclooxygenase II helps to maintain an enhanced response to catecholamines for an extended period, spanning several days. The successful reinstatement of vascular responsiveness to vasopressors might be intricately linked to the degree of disparity in the activities of NF-κB and the G-GRα complex [19]. Administering moderate doses of hydrocortisone in septic shock patients led to improved capillary density and blood flow, effects that were observed as soon as one hour after treatment [20]. The improvement in microcirculation is believed to stem from heightened activity of the endothelial form of nitric oxide synthase, facilitated by a mechanism involving the mitogen-activated protein kinase/Akt pathway [21]. Moreover, corticosteroids might play a role in restoring the inherent fluctuations present in the cardiovascular system [22].

- Organ restoration and ICU length of stay

In septic shock patients, glucocorticoids were found to suppress the release of tumor necrosis factors from vascular and smooth muscle tissues [23]. Additionally, by the fifth day of administering hydrocortisone, there was a total inhibition of NF-κB activity in peripheral mononuclear cells. Corticosteroids have proven successful in diminishing renal iNOS activity after endotoxemia, aiding in the prevention of hypoxic injury to the kidney cortex, improving oxygen supply to the kidneys, and ultimately normalizing the use of oxygen by the kidneys [24]. Similarly, corticosteroids have been shown to enhance the permeability of the glomerular endothelium in septic shock patients [25]. Corticosteroids have a beneficial effect on the cardiovascular, pulmonary, hepatic, and renal systems. Consequently, patients undergoing corticosteroid treatment may rapidly transition off vasopressor support and mechanical ventilation, facilitating their earlier release from the intensive care unit [26].

Although so much evidence about the good side of corticosteroids is to be found in the literature, their administration in sepsis or septic shock is not yet convincing. The latest guidelines from the Surviving Sepsis Campaign for the use of corticosteroids in adults with septic shock recommend intravenous hydrocortisone at a daily dosage of 200 mg. This dosage is typically delivered as 50 mg, given intravenously every 6 h or through a continuous infusion. It is recommended to start this corticosteroid treatment at least

4 h after the initiation of norepinephrine or epinephrine, at a dose of ≥0.25 mcg/kg/min in adults. This is categorized as a weak recommendation with a moderate quality of evidence [2].

This review's objective is to present the current evidence, based on thoroughly conducted and extensive studies, on the use of corticosteroids in sepsis and septic shock with the scope of creating a clearer image of their use in septic and septic shock patients.

3. Materials and Method

For the study research, the PubMed database was searched using the MesH terms "sepsis", septic shock", and "corticosteroids". The search retrieved 26,577 results, which, after applying filters such as being published in the last five years, randomized controlled studies, exclusion of books, abstracts, and no free full text available, written in a language other than English, as well as reviews and meta-analyses, were narrowed to 12 multicentric, randomized control studies. The flow chart below presents the study's search process (Figure 1).

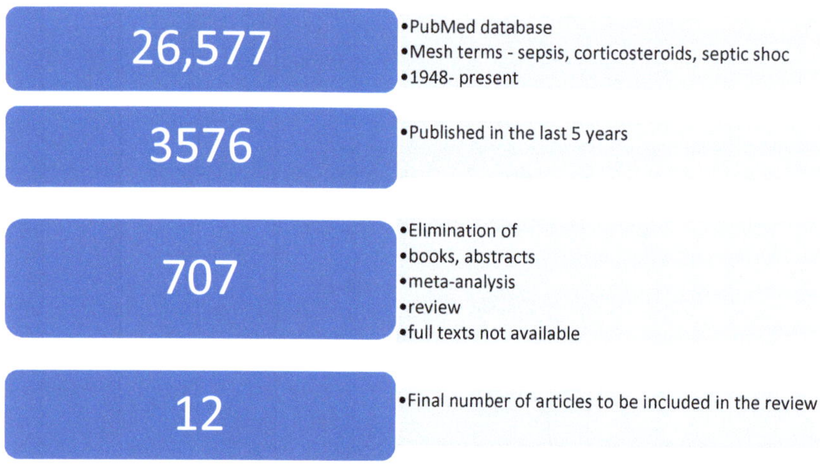

Figure 1. Flow chart of the study's search process.

The data processed from the final included articles were as follows: if the study was multicentric, the number of patients included, the intervention, the main results, and the conclusions of the study.

4. Results

Our findings are presented in Table 1.

The dosages of the corticosteroid treatments and their duration from the selected studies are presented below in Table 2.

Table 1. The findings of this study.

Study/Year Published	Multicentric Yes/No	No. of PTS	Intervention	Results	Conclusion
Resuscitation With Vitamin C, Hydrocortisone, and Thiamin in Children With Septic Shock: A Multicenter Randomized Pilot Study [27] 2024	YES	60	Patients assigned to the group receiving vitamin C, hydrocortisone, and thiamin were administered 30 mg/kg per dose of vitamin C (up to a maximum of 1500 mg per dose as sodium ascorbate, sourced from biological therapies) intravenously every 6 h, 1 mg/kg per dose of hydrocortisone (with a maximum of 50 mg per dose) intravenously every 6 h, and 4 mg/kg per dose of thiamin (up to a maximum of 200 mg per dose) intravenously every 12 h, starting immediately after they were randomized. The treatment was continued for 7 days or until any of the following occurred: the shock was resolved, the patient died, the patient was discharged from the pediatric intensive care unit (PICU), or if any major adverse events attributed to the treatment were observed.	By the 28th day, participants in the intervention group had a median of 20.0 organ dysfunction-free days, compared to 21.0 days in the group receiving standard care. The median duration without the need for inotrope support within the first 7 days was 6.3 days for those in the intervention group and 5.9 days for those receiving standard care. Within the first 28 days post randomization, 15% (4 out of 27) of patients in the intervention group passed away, as opposed to 6% (2 out of 33) in the standard care group. The median stay in the oediatric intensive care unit (PICU) was 5.3 days for the intervention group and 6.9 days for the standard care group.	A pragmatic trial on high-dose vitamin C, hydrocortisone, and thiamin in children with septic shock admitted to PICU is feasible.
Evaluation of hydrocortisone, vitamin C, and thiamine for the treatment of septic shock: a randomized controlled trial (The HYVITS Trial) [28] 2023	YES	106	The three-drug treatment involved administering 1.5 g of intravenous vitamin C every 6 h for 4 days or until the patient was discharged from the ICU, whichever came first, 50 mg of hydrocortisone every 6 h for 7 days or until ICU discharge, followed by a gradual reduction over 3 days, and 200 mg of intravenous thiamine every 12 h for 4 days or until ICU discharge, whichever occurred sooner. This was in comparison to providing only standard care to patients experiencing septic shock.	No significant statistical differences were observed between the group receiving triple therapy and the control group in terms of: Mortality within the hospital at 60 days. Rates of discharge from the hospital. The length of vasopressor use among those who survived. The amount of time spent on mechanical ventilation. The variation in SOFA (sequential organ failure assessment) score at 72 h. The requirement for renal replacement therapy was comparable between the two groups. Outcomes related to safety. However, a higher incidence of failure to wean off mechanical ventilation was noted in the triple therapy group (80% compared to 64.1%).	In patients with septic shock who needed vasopressor support, the combined use of hydrocortisone, vitamin C, and thiamine did not lead to a reduction in 60-day in-hospital mortality, nor did it shorten the length of vasopressor use or lower SOFA scores at 72 h.

Table 1. *Cont.*

Study/Year Published	Multicentric Yes/No	No. of PTS	Intervention	Results	Conclusion
Effects of hydrocortisone combined with vitamin C and vitamin B1 versus hydrocortisone alone on micro-circulation in septic shock patients: A pilot study [29] 2023	NO (pilot study)	27	The treatment group received a combination of hydrocortisone, vitamin C, and vitamin B1, in addition to standard care. The control group was administered hydrocortisone as a standalone treatment, alongside standard care.	Between the treatment and control groups, no statistically significant differences were observed in: sPVD (skin perfusion video densitometry) at the outset. Baseline measures of sPPV (skin perfusion pressure variability), sTVD (skin total vessel density), or MFI (microcirculatory flow index). Time to peak (TTP) or mean transit time (MTT) 24 h post treatment in the treatment group versus the control group. However, after adjusting for baseline perfusion index (PI) and renal blood flow (RBF), both PI and RBF were significantly greater in the treatment group than in the control group 24 h after receiving treatment. Lactate levels were notably lower in the treatment group compared to the control group 24 h post treatment. No significant variance was found in the doses of norepinephrine administered at baseline, 4 h, and 24 h post treatment between the two groups. The treatment group exhibited significantly higher values for sPPV, sTVD, and MFI compared to the control group 24 h after treatment.	The combination therapy was significantly more effective at enhancing microcirculation in patients with septic shock than hydrocortisone alone.
Corticotropin-stimulated steroid profiles to predict shock development and mortality in sepsis: From the HYPRESS study [30] 2022	YES	206	Corticotropin test for all included patients.	In healthy participants, the response to corticotropin stimulation varied widely, showing a highly dynamic reaction across all analyzed steroids and their precursors. When comparing the sepsis group to healthy individuals, baseline corticosterone levels were similar, but cortisone levels were notably lower in the sepsis group. Following corticotropin stimulation, sepsis patients exhibited the same levels of the precursor 17-OH progesterone and significantly elevated levels of 11-desoxycorticosterone, 11-desoxycortisol, and cortisol compared to healthy subjects. However, the rise in corticosterone was significantly less in sepsis patients than in healthy subjects. Comparing healthy individuals with patients having severe sepsis revealed that 50% of the sepsis patients (those not experiencing shock) had a below normal increase in corticosterone. Analysis of outcomes, specifically in-hospital mortality within the placebo group, indicated that sepsis patients who did not survive had significantly smaller increases in corticosterone than those who did survive, pointing to a more severe disruption in mineralocorticoid metabolism among the non-survivors.	Steroid profiling indicated that the mineralocorticoid pathway was often more affected than the glucocorticoid pathway in the body's stress response to sepsis. Patients exhibiting higher levels of glucocorticoids relative to mineralocorticoids following corticotropin stimulation had a higher likelihood of progressing to septic shock and succumbing in the hospital. The cortisol-to-corticosterone ratio post corticotropin stimulation can serve as a predictor for critical outcomes like the onset of shock and death in sepsis cases. This ratio may also guide the application of hydrocortisone treatment.

Table 1. *Cont.*

Study/Year Published	Multicentric Yes/No	No. of PTS	Intervention	Results	Conclusion
Early administration of hydrocortisone, vitamin C, and thiamine in adult patients with septic shock: a randomized controlled clinical trial [31]. 2022	NO	408	Patients suffering from septic shock were mainly managed with an aggressive fluid resuscitation strategy, appropriate antibiotics, and vasoactive medications, following the guidelines of the Surviving Sepsis Campaign. Norepinephrine was chosen as the initial vasoactive medication. The treatment group received a regimen of hydrocortisone (200 mg daily), vitamin C (2 g every 6 h), and thiamine (200 mg every 12 h) for 5 days or until patients were discharged from the ICU. Conversely, the placebo group was given an equivalent volume of 0.9% saline from a placebo container, adhering to the same administration schedule.	In the intention-to-treat (ITT) analysis, there was no significant difference in 90-day survival between the groups. In the per-protocol (PP) analysis, 90-day survival also showed no significant difference between the groups. Within the ITT population, there was no noticeable difference in 28-day mortality, or ICU mortality, between those receiving the intervention and those receiving a placebo. For the PP population, there was no significant variance in 28-day all-cause mortality, mortality upon ICU discharge, or hospital discharge between the intervention and placebo groups. The ITT analysis revealed that the rate of shock reversal was comparable between the intervention and placebo groups. There was no statistically significant difference observed in the 72 h delta SOFA score, nor in the 28-day cumulative days free from the ICU, vasopressors, or ventilatory support between the two groups. No notable differences were found regarding the length of stay (LOS) in the ICU or the hospital between the intervention and placebo groups. In the PP analysis, the proportion of patients experiencing shock reversal was alike in both the intervention and placebo groups. There was no statistical significance found in the 28-day cumulative ICU-free days between the groups. Regarding adverse events within the safety population, the most frequently occurring serious adverse events were severe hypernatremia and fluid overload. A disturbance in blood glucose levels was noted in 27 patients within the intervention group	In individuals with septic shock, the combination of hydrocortisone, vitamin C, and thiamine did not demonstrate a reduction in 90-day mortality when compared to a placebo. Based on these findings, the regular application of this therapy mix for adult patients experiencing septic shock is not advocated.

Table 1. *Cont.*

Study/Year Published	Multicentric Yes/No	No. of PTS	Intervention	Results	Conclusion
Effect of 12 mg vs. 6 mg of Dexamethasone on the Number of Days Alive Without Life Support in Adults With COVID-19 and Severe Hypoxemia [32] 2021	YES	1000	Two study groups: 1. Received 6 mg of dexamethasone. 2. Received 12 mg of dexamethasone.	Participants in the 12 mg dexamethasone group experienced a median of 22.0 days alive without the need for life support, compared to 20.5 days in the 6 mg dexamethasone group. The 28-day mortality rate was 27.1% for those receiving 12 mg of dexamethasone, versus 32.3% for the 6 mg dexamethasone group. At 90 days, the mortality rate was 32.0% in the group given 12 mg of dexamethasone, against 37.7% in the group receiving 6 mg of dexamethasone. Major adverse effects, such as septic shock and invasive fungal infections, were reported in 11.3% of patients in the 12 mg dexamethasone group compared to 13.4% in the 6 mg dexamethasone group.	In patients with COVID-19 experiencing severe hypoxemia, administering 12 mg/day of dexamethasone as opposed to 6 mg/day did not significantly increase the number of days alive without life support within a 28-day period.
External corroboration that corticosteroids may be harmful to septic shock endotype A patients [33] 2021	Yes	97	Utilizing transcriptomic information from the VANISH trial, research subjects were categorized into pediatric septic shock endotypes A or B. This classification was based on the comparison of each subject's gene expression profiles with the reference profiles for endotypes A and B [34]. The study aimed to explore the theory that hydrocortisone therapy correlates with higher mortality rates in patients classified under endotype A.	At the start of the study, there were no notable differences between the two endotypes. Patients with endotype A who received hydrocortisone had a mortality rate of 46%, in contrast to a 22% mortality rate for those with endotype A who were given a placebo. For patients with endotype B, the mortality rate was 22% for those treated with hydrocortisone and 19% for those receiving a placebo. The comparison between groups indicated that the likelihood of death was three times higher for endotype A patients treated with hydrocortisone compared to endotype B patients who received a placebo. No significant difference in mortality risk was observed between endotype B patients on placebo and either endotype A patients on placebo or endotype B patients on hydrocortisone.	This investigative analysis offers additional support for the possibility that exposure to corticosteroids could be linked to a higher mortality rate in patients with septic shock endotype A.

Table 1. *Cont.*

Study/Year Published	Multicentric Yes/No	No. of PTS	Intervention	Results	Conclusion
Effect of Ascorbic Acid, Corticosteroids, and Thiamine on Organ Injury in Septic Shock [35] 2020	YES	200	Participants were randomly allocated to either receive an intravenous combination of ascorbic acid (1500 mg), hydrocortisone (50 mg), and thiamine (100 mg) at 6 h intervals for four days or a placebo in equivalent volumes administered at the same intervals.	Over the 72 h following enrollment, no statistically significant difference was observed between the treatment and placebo groups regarding the change in SOFA score over time. There was no statistically significant difference in the occurrence of kidney failure between those receiving the intervention and those given a placebo, nor was there a difference in 30-day mortality rate. The most frequently reported serious adverse events included hyperglycemia (12 cases in the intervention group versus 7 in the placebo group), hypernatremia (11 in the intervention group versus 7 in the placebo group), and the onset of new hospital-acquired infections (13 in the intervention group versus 12 in the placebo group).	For patients experiencing septic shock, the use of a treatment regimen combining ascorbic acid, corticosteroids, and thiamine, as opposed to a placebo, did not lead to a statistically significant decrease in the SOFA score within the initial 72 h post enrollment. Based on these findings, the regular application of this combination therapy for septic shock patients is not endorsed.
Effect of Vitamin C, Hydrocortisone, and Thiamine vs. Hydrocortisone Alone on Time Alive and Free of Vasopressor Support Among Patients With Septic Shock: The VITAMINS Randomized Clinical Trial [36] 2020	YES	216	Participants were allocated randomly to either the treatment group, which received an intravenous combination of vitamin C (1.5 g every 6 h), hydrocortisone (50 mg every 6 h), and thiamine (200 mg every 12 h), or to the control group, which was administered only intravenous hydrocortisone (50 mg every 6 h). This regimen was continued until the resolution of shock or for a maximum of 10 days.	The duration of survival without the need for vasopressors by the 7th day was 122.1 h for the treatment group and 124.6 h for the control group. Among the 10 secondary outcomes predefined for examination, 9 exhibited no statistically meaningful differences. The mortality rate at 90 days stood at 28.6% (30 out of 105) in the group receiving the intervention and 24.5% (25 out of 102) in the control group. No incidents of serious adverse effects were documented.	For patients experiencing septic shock, administering a combination of intravenous vitamin C, hydrocortisone, and thiamine did not markedly extend the period they remained alive without the need for vasopressors over 7 days, in comparison to those treated with only intravenous hydrocortisone. This outcome indicates that the combined treatment regimen does not facilitate a quicker recovery from septic shock than the use of intravenous hydrocortisone by itself.

Table 1. *Cont.*

Study/Year Published	Multicentric Yes/No	No. of PTS	Intervention	Results	Conclusion
The cost-effectiveness of adjunctive corticosteroids for patients with septic shock [37]. 2020	YES	1513	The quality of life related to health 6 months post treatment was assessed using the EuroQoL 5-dimension 5-level questionnaire. Information on the utilization of hospital resources and associated expenses was gathered by integrating the ADRENAL dataset with governmental health administration databases. Measured clinical outcomes encompassed mortality, health-related quality of life, and the addition of quality-adjusted life years. Economic outcomes considered were the use of hospital resources, associated costs, and cost-effectiveness from the viewpoint of the healthcare payer.	No significant difference was observed in mortality rate or overall hospital expenses between the group receiving hydrocortisone and that given a placebo. The additional cost associated with hydrocortisone, for each quality-adjusted life-year gained, amounted to A $1,254,078. In the case of female patients, hydrocortisone proved to be cost-effective in 46.2% of bootstrapped replicates, while for male patients, it was deemed cost-effective in only 2.7% of these replicates.	The use of hydrocortisone as an additional treatment did not significantly impact long-term mortality, health-related quality of life, the utilization of healthcare resources, or costs, making it improbable to be considered cost-effective.
Hydrocortisone Compared with Placebo in Patients with Septic Shock Satisfying the Sepsis-3 Diagnostic Criteria and APROCCHSS Study Inclusion Criteria: A Post Hoc Analysis of the ADRENAL Trial [38] 2019	YES	2855	A subsequent analysis of the ADRENAL database was conducted to pinpoint patient groups that fulfilled the Sepsis-3 criteria for septic shock (termed ADRENAL–Sepsis-3) or the inclusion criteria for APROCCHSS (referred to as ADRENAL–APROCCHSS).	In patient groups from the ADRENAL study who matched the criteria for either Sepsis-3 or APROCCHSS, a greater mortality rate was observed at 90 days. However, hydrocortisone treatment did not significantly reduce mortality compared to placebo. The three groups showed similar results in several secondary outcomes, including a quicker reversal of shock with hydrocortisone, the frequency of mechanical ventilation resumption, the number of days spent alive and outside the hospital, and the occurrence of new bacteremia or fungemia. In both patient subsets receiving hydrocortisone, there was an increased incidence of shock recurrence, yet the impact on rates of blood transfusion did not differ by treatment. For patients who conformed to the Sepsis-3 criteria, those treated with hydrocortisone experienced a decrease in mortality at 28 days, an extension in the duration of being alive without the need for mechanical ventilation or renal replacement therapy, and more days spent alive and outside of the ICU.	Among participants in the ADRENAL trial who met the criteria for Sepsis-3 or APROCCHSS, administering hydrocortisone via continuous infusion did not significantly reduce the 90-day mortality rate compared to a placebo in cases of septic shock.

Table 1. *Cont.*

Study/Year Published	Multicentric Yes/No	No. of PTS	Intervention	Results	Conclusion
Transcriptomic Signatures in Sepsis and a Differential Response to Steroids. From the VANISH Randomized Trial [39] 2019	YES	176	A post hoc analysis was performed of a double-blind, randomized clinical trial in septic shock (VANISH [Vasopressin vs. Norepinephrine as Initial Therapy in Septic Shock]) [40]. Participants were enrolled within 6 h following the onset of shock. They were then randomly assigned to receive either norepinephrine or vasopressin, which was followed by the administration of hydrocortisone or a placebo. Comprehensive gene expression analysis was carried out across the genome, and the SRS endotype was identified through a pre-established model that utilizes seven distinct genes for discrimination.	No significant correlation was found between the SRS group and the type of vasopressor used; however, a significant interaction was observed between the allocation to hydrocortisone or placebo and the SRS endotype. Specifically, the use of hydrocortisone was linked to higher mortality rates in individuals exhibiting the SRS2 phenotype.	The gene expression pattern observed at the beginning of septic shock was related to how patients responded to corticosteroids. Individuals characterized by the immuno-competent SRS2 endotype experienced a notably higher mortality rate when treated with corticosteroids versus a placebo.

Table 2. Corticosteroid treatment in the selected studies—dose and duration of treatment.

Study/Year Published	Corticosteroid Dosage	Duration of Corticosteroid Treatment
Resuscitation With Vitamin C, Hydrocortisone, and Thiamin in Children With Septic Shock: A Multicenter Randomized Pilot Study [27] 2024	1 mg/kg per dose of hydrocortisone (with a maximum of 50 mg per dose) intravenously every 6 h immediately after patients were randomized in combination with vitamin C and thiamine.	The treatment was continued for 7 days or until any of the following occurred: the shock was resolved, the patient died, the patient was discharged from the pediatric intensive care unit (PICU), or if any major adverse events attributed to the treatment were observed.
Evaluation of hydrocortisone, vitamin C, and thiamine for the treatment of septic shock: a randomized controlled trial (The HYVITS trial) [28] 2023	50 mg of hydrocortisone every 6 h over 3 days, in combination with vitamin C and thyamine.	For 7 days or until ICU discharge, followed by gradual reduction.
Effects of hydrocortisone combined with vitamin C and vitamin B1 versus hydrocortisone alone on microcirculation in septic shock patients: A pilot study [29] 2023	Hydrocortisone (200 mg) was continuously pumped intravenously for 24 h in combination with vitamin C and thiamine or as a standalone treatment.	The infusion time was 30–60 min, and the interval was 6 h; 4 doses were used nt.
Early administration of hydrocortisone, vitamin C, and thiamine in adult patients with septic shock: a randomized controlled clinical trial [31]. 2022	Hydrocortisone (200 mg daily) along with vitamin C and thiamine vs. placebo.	For 5 days or until patients were discharged from the ICU.
Effect of 12 mg vs. 6 mg of Dexamethasone on the Number of Days Alive Without Life Support in Adults With COVID-19 and Severe Hypoxemia [32] 2021	Two study groups 1. Received 6 mg of dexamethasone. 2. Received 12 mg of dexamethasone.	
Effect of Ascorbic Acid, Corticosteroids, and Thiamine on Organ Injury in Septic Shock [35] 2020	Intravenous combination of ascorbic acid, hydrocortisone (50 mg), and thiamine at 6 h intervals vs. placebo.	4 days
Effect of Vitamin C, Hydrocortisone, and Thiamine vs. Hydrocortisone Alone on Time Alive and Free of Vasopressor Support Among Patients With Septic Shock: The VITAMINS Randomized Clinical Trial [36] 2020	Intravenous combination of vitamin C hydrocortisone (50 mg every 6 h), and thiamine vs. control group intravenous hydrocortisone (50 mg every 6 h).	This regimen continued until the resolution of shock or for a maximum of 10 days.

5. Discussion

While corticosteroids are considered for use in certain cases of sepsis and septic shock, their administration is treated with caution due to the possibility of adverse effects, increased risk of exacerbating infections, and unpredictable reactions among individual patients. The decision to employ corticosteroids in treating sepsis requires a nuanced assessment of the potential advantages and dangers for each patient. Corticosteroids, which are potent anti-inflammatory agents, can adjust the immune system's response. Given the severity of sepsis—a critical, life-endangering reaction to infection that may cause tissue harm, organ failure, and mortality—the application of corticosteroids has sparked significant discussion and investigation.

Some of the reasons for which corticosteroids are included in the last line of sepsis, aspects which constitute the disadvantages of their use in septic shock, include:

- Corticosteroids can suppress the immune system. In the setting of sepsis, where the body is fighting a severe infection, further suppression of the immune response may be counterproductive, potentially allowing the underlying infection to worsen [41].
- Due to their immunosuppressive effects, corticosteroids can increase the risk of secondary infections, which can complicate or exacerbate the patient's condition [42].
- Corticosteroids can increase blood sugar levels, which could complicate the management of septic patients, especially those with diabetes or those who develop stress-induced hyperglycemia because of their critical illness. One detail is very important about this statement, namely that most of these glucose level impairments related to corticosteroids in septic patients appear mainly when boluses are administered [43].
- They can also cause fluid retention and electrolyte imbalance, potentially exacerbating sepsis-induced organ dysfunction, such as acute kidney injury [44].

6. Different Combinations of Corticosteroids

Despite the theoretical benefits of corticosteroids, vitamin C, and thiamine combination therapy, including anti-inflammatory and antioxidant properties and the potential to preserve endothelial function, the trials found no significant difference in primary outcomes like the change in the SOFA score at 72 h or secondary outcomes such as kidney failure, 30-day mortality, ventilator-free days, and ICU-free days between the intervention and placebo groups. However, the intervention group did have a statistically significant increase in the number of shock-free days compared to the placebo group [31,35,36].

These trials highlight the ongoing debate regarding the utility of ascorbic acid, hydrocortisone, and thiamine in septic shock treatment. Despite initial enthusiasm based on smaller, single-center studies, larger multicenter trials have not demonstrated a significant benefit in reducing organ failure, mortality, or other critical outcomes with this triple therapy therapy. The only consistent finding across trials has been a reduction in the duration of vasopressor support, suggesting a potential role for triple therapy in specific patient subgroups or phenotypes, rather than a universal treatment for all septic shock patients [27–29].

The study by Marik et al. published in *Chest* in June 2017 investigated the effects of a combination treatment involving hydrocortisone, vitamin C, and thiamine on patients with severe sepsis and septic shock. This retrospective before–after study involved comparing the outcomes of 47 consecutive septic patients treated with this combination therapy against 47 historical control patients who received standard of care.

The main findings of the study were significant. The treatment group that received the combination therapy showed a remarkable reduction in mortality compared to the control group (8.5% vs. 40.4%). Furthermore, the patients treated with the combination therapy also demonstrated a reduction in the duration of vasopressor administration, indicating an improvement in vascular responsiveness.

The authors concluded that the administration of hydrocortisone, vitamin C, and thiamine appears to be safe and could lead to a significant reduction in mortality in patients with severe sepsis and septic shock. However, they also suggested that these findings should be validated by randomized controlled trials to confirm the efficacy and safety of this treatment approach [45].

These results underscore the complexity of treating septic shock and the importance of ongoing research to refine and discover effective therapies. Current evidence does not support the widespread use of triple therapy to improve mortality or organ dysfunction in septic shock [27–29,31,35,36]. Future trials may provide further insight into whether there are specific patient populations who could benefit from this therapy.

7. Genomics in Sepsis Treatment

An emerging source of significant results in corticoid use in sepsis appears to be genomic subtyping and treatment personalization [33,39]. Wong HR et al. focused on a crucial component of personalized medicine in treating sepsis, especially in cases of septic shock. Their study examined the impact of corticosteroids on patients identified with a particular endotype of septic shock, known as endotype A,. This focus emphasizes the growing recognition and significance of genomics in the treatment of sepsis, showcasing how personalized healthcare strategies are becoming increasingly important in managing this complex condition [33].

The study by Wong and colleagues marks an important advancement in the treatment of sepsis, offering critical perspectives on the utilization of genomic data to enhance patient results. It underscores the shift toward a personalized treatment paradigm in sepsis management, emphasizing the importance of considering an individual's genetic profile to refine therapeutic approaches. This investigation illuminates both the promising advantages of personalized medicine and the obstacles and factors that need attention as genomics increasingly influences clinical decisions in sepsis and beyond [33].

The research presented by Antcliffe and colleagues represents a significant step forward in the application of genomic technologies in sepsis treatment. By demonstrating how transcriptomic profiles can differentiate responses to corticosteroids among sepsis patients, this work paves the way for more personalized, effective, and safe treatment strategies [39]. This research utilized data from the VANISH (Vasopressin vs. Norepinephrine as Initial Therapy in Septic Shock) randomized trial to examine how genomic technologies, especially transcriptomics, might redefine sepsis management, with a focus on corticosteroid administration [40]. The research uncovered distinct transcriptomic patterns in sepsis patients, linking these patterns to variations in how individuals respond to corticosteroid treatments. These findings provide valuable insight into the biological mechanisms and pathways that are activated in sepsis and their interaction with treatment methods.

The study highlights the capacity of transcriptomics to personalize corticosteroid therapy for sepsis patients. Identifying patients likely to respond favorably to corticosteroids through their transcriptomic profile allows for more precise medical decisions, potentially enhancing patient outcomes and minimizing negative side effects [39].

Transcriptomic analyses shed light on the specific ways in which corticosteroids affect patients with septic shock, offering a clearer understanding of the drugs' biological actions. Such insight could guide the discovery of biomarkers for treatment prediction and the development of novel treatment avenues.

8. Conclusions

Recent advancements underscore a move toward individualized medicine in managing sepsis, highlighting the potential of genomic and transcriptomic analyses to inform more customized treatment approaches. Such insight could significantly enhance patient outcomes by identifying those who are likely to benefit from corticosteroid therapy and those who may face detrimental effects. However, integrating genomic personalization into clinical practice faces considerable challenges, including the need for rapid, accurate genomic assessments and the creation of clinical protocols to effectively use genomic information.

There is a pressing call for additional studies to validate and expand the use of genomic and transcriptomic data in guiding therapeutic decisions for sepsis and septic shock. This includes exploring further genomic indicators and omics techniques to improve the customization of care and patient outcomes.

In essence, while corticosteroids are crucial in treating sepsis and septic shock, their use requires careful consideration of potential benefits against risks. The introduction of genomic technologies offers the promise of treatment plans tailored to individual patients, yet fully realizing this potential necessitates ongoing research and overcoming current barriers to implementation.

Funding: This research received no funding.

Data Availability Statement: Data sharing is not applicable.

Conflicts of Interest: The author declares no conflicts of interest.

References

1. Singer, M.; Deutschman, C.S.; Seymour, C.W.; Shankar-Hari, M.; Annane, D.; Bauer, M.; Bellomo, R.; Bernard, G.R.; Chiche, J.-D.; Coopersmith, C.M.; et al. The Third International Consensus Definitions for Sepsis and Septic Shock (Sepsis-3). *JAMA* **2016**, *315*, 801–810. [CrossRef] [PubMed]
2. Evans, L.; Rhodes, A.; Alhazzani, W.; Antonelli, M.; Coopersmith, C.M.; French, C.; Machado, F.R.; Mcintyre, L.; Ostermann, M.; Prescott, H.C.; et al. Surviving sepsis campaign: International guidelines for management of sepsis and septic shock 2021. *Intensive Care Med.* **2021**, *47*, 1181–1247. [CrossRef] [PubMed]
3. Chiu, C.; Legrand, M. Epidemiology of sepsis and septic shock. *Curr. Opin. Anaesthesiol.* **2021**, *34*, 71–76. [CrossRef] [PubMed]
4. Piechota, M.; Banach, M.; Irzmanski, R.; Barylski, M.; Piechota-Urbanska, M.; Kowalski, J.; Pawlicki, L. Plasma endothelin-1 levels in septic patients. *J. Intensive Care Med.* **2007**, *22*, 232–239. [CrossRef] [PubMed]
5. Ince, C. The microcirculation is the motor of sepsis. *Crit. Care* **2005**, *9*, S13–S19. [CrossRef] [PubMed]
6. Belousoviene, E.; Kiudulaite, I.; Pilvinis, V.; Pranskunas, A. Links between Endothelial Glycocalyx Changes and Microcirculatory Parameters in Septic Patients. *Life* **2021**, *11*, 790. [CrossRef]
7. Hahn, E.O.; Houser, H.B.; Rammelkamp, C.H.; Denny, F.W.; Wannamaker, L.W. Effect of cortisone on acute streptococcal infections and post-streptococcal complications. Effect of cortisone on acute streptococcal infections and post-streptococcal complications. *J. Clin. Investig.* **1951**, *30*, 274–281. [CrossRef] [PubMed]
8. Britt, R.C.; Devine, A.; Swallen, K.C.; Weireter, L.J.; Collins, J.N.; Cole, F.J.; Britt, L.D. Corticosteroid Use in the Intensive Care Unit: At What Cost? *Arch. Surg.* **2006**, *141*, 145–149. [CrossRef] [PubMed]
9. Tracey, K.J. Physiology and immunology of the cholinergic antiinflammatory pathway. *J. Clin. Investig.* **2007**, *117*, 289–296. [CrossRef]
10. Barnes, P.J.; Karin, M. Nuclear factor-kappaB: A pivotal transcription factor in chronic inflammatory diseases. *N. Engl. J. Med.* **1997**, *336*, 1066–1071. [CrossRef]
11. Kashiwabara, M.; Miyashita, M.; Nomura, T.; Makino, H.; Matsutani, T.; Kim, C.; Takeda, S.; Yamashita, K.; Chaudry, I.H.; Tajiri, T. Surgical trauma-induced adrenal insufficiency is associated with postoperative inflammatory responses. *J. Nippon. Med. Sch.* **2007**, *74*, 274–283. [CrossRef] [PubMed]
12. Prigent, H.; Maxime, V.; Annane, D. Science review: Mechanisms of impaired adrenal function in sepsis and molecular actions of glucocorticoids. *Crit. Care* **2004**, *8*, 243–252. [CrossRef] [PubMed]
13. Moraes, L.A.; Paul-Clark, M.J.; Rickman, A.; Flower, R.J.; Goulding, N.J.; Perretti, M. Ligand-specific glucocorticoid receptor activation in human platelets. *Blood* **2005**, *106*, 4167–4175. [CrossRef] [PubMed]
14. Rhen, T.; Cidlowski, J.A. Antiinflammatory action of glucocorticoids: New mechanisms for old drugs. *N. Engl. J. Med.* **2005**, *353*, 1711–1723. [CrossRef] [PubMed]
15. Ehrchen, J.; Steinmüller, L.; Barczyk, K.; Tenbrock, K.; Nacken, W.; Eisenacher, M. Glucocorticoids induce differentiation of a specifically activated, anti-inflammatory subtype of human monocytes. *Blood* **2007**, *109*, 1265–1274. [CrossRef] [PubMed]
16. Barczyk, K.; Ehrchen, J.; Tenbrock, K.; Ahlmann, M.; Kneidl, J.; Viemann, D.; Roth, J. Glucocorticoids promote survival of anti-inflammatory macrophages via stimulation of adenosine receptor A3. *Blood* **2010**, *116*, 446–455. [CrossRef]
17. Varga, G.; Ehrchen, J.; Tsianakas, A.; Tenbrock, K.; Rattenholl, A.; Seeliger, S.; Mack, M.; Roth, J.; Sunderkoetter, C. Glucocorticoids induce an activated, anti-inflammatory monocyte subset in mice that resembles myeloid-derived suppressor cells. *J. Leukoc. Biol.* **2008**, *84*, 644–650. [CrossRef]
18. Boyer, A.; Chadda, K.; Salah, A.; Annane, D. Glucocorticoid treatment in patients with septic shock: Effects on vasopressor use and mortality. *Int. J. Clin. Pharmacol. Ther.* **2006**, *44*, 309–318. [CrossRef]
19. Bianca, R.D.D.V.; Lippolis, L.; Autore, G.; Popolo, A.; Marzocco, S.; Sorrentino, L.; Pinto, A.; Sorrentino, R. Dexamethasone improves vascular hyporeactivity induced by LPS in vivo by modulating ATP-sensitive potassium channels activity. *Br. J. Pharmacol.* **2003**, *140*, 91–96. [CrossRef]

20. Büchele, G.L.; Silva, E.; Ospina-Tascón, G.A.; Vincent, J.L.; De Backer, D. Effects of hydrocortisone on microcirculatory alterations in patients with septic shock. *Crit. Care Med.* **2009**, *37*, 1341–1347. [CrossRef]
21. Hafezi-Moghadam, A.; Simoncini, T.; Yang, Z.; Limbourg, F.P.; Plumier, J.-C.; Rebsamen, M.C.; Hsieh, C.-M.; Chui, D.-S.; Thomas, K.L.; Prorock, A.J.; et al. Acute cardiovascular protective effects of corticosteroids are mediated by non-transcriptional activation of endothelial nitric oxide synthase. *Nat. Med.* **2002**, *8*, 473–479. [CrossRef] [PubMed]
22. Aboab, J.; Polito, A.; Orlikowski, D.; Sharshar, T.; Castel, M.; Annane, D. Hydrocortisone effects on cardiovascular variability in septic shock: A spectral analysis approach. *Crit. Care Med.* **2008**, *36*, 1481–1486. [CrossRef] [PubMed]
23. Newman, W.H.; Zhang, L.M.; Leeper-Woodford, S.K.; Shaker, I.J.; Erceg, S.K.; Castresana, M.R. Inhibition of release of tumor necrosis factor-alpha from human vascular tissue and smooth muscle cells by glucocorticoids. *Crit. Care Med.* **1997**, *25*, 519–522. [CrossRef] [PubMed]
24. Johannes, T.; Mik, E.G.; Klingel, K.; Dieterich, H.J.; Unertl, K.E.; Ince, C. Low-dose dexamethasone-supplemented fluid resuscitation reverses endotoxin-induced acute renal failure and prevents cortical microvascular hypoxia. *Shock* **2009**, *31*, 521–528. [CrossRef] [PubMed]
25. Rinaldi, S.; Adembri, C.; Grechi, S.; De Gaudio, R. Low-dose hydrocortisone during severe sepsis: Effects on microalbuminuria. *Crit. Care Med.* **2006**, *34*, 2334–2339. [CrossRef] [PubMed]
26. Annane, D. Corticosteroids for severe sepsis: An evidence-based guide for physicians. *Ann. Intensive Care* **2011**, *13*, 7. [CrossRef] [PubMed]
27. Schlapbach, L.J.; Raman, S.; Buckley, D.; George, S.; King, M.; Ridolfi, R.; Harley, A.; Cree, M.; Long, D.; Erickson, S.; et al. Resuscitation With Vitamin C, Hydrocortisone, and Thiamin in Children With Septic Shock: A Multicenter Randomized Pilot Study. *Pediatr. Crit. Care Med.* **2024**, *25*, 159–170. [CrossRef] [PubMed]
28. Mohamed, A.; Abdelaty, M.; Saad, M.O.; Shible, A.; Mitwally, H.; Akkari, A.-R.; Elbuzidi, A.; Bintaher, A.; Hashim, A.; Abdelrahim, M.; et al. Evaluation of hydrocortisone, vitamin c, and thiamine for the treatment of septic shock: A randomized controlled trial (THE HYVITS TRIAL). *Shock* **2023**, *1*, 697–701. [CrossRef] [PubMed]
29. Wang, J.; Song, Q.; Yang, S.; Wang, H.; Meng, S.; Huang, L.; Li, Q.; Xu, J.; Xie, J.; Huang, Y. Effects of hydrocortisone combined with vitamin C and vitamin B1 versus hydrocortisone alone on microcirculation in septic shock patients: A pilot study. *Clin. Hemorheol. Microcirc.* **2023**, *84*, 111–123. [CrossRef]
30. Briegel, J.; Möhnle, P.; Keh, D.; Lindner, J.M.; Vetter, A.C.; Bogatsch, H.; Lange, D.; Frank, S.; Hinske, L.C.; Annane, D.; et al. SepNet Critical Care Trials. Corticotropin-stimulated steroid profiles to predict shock development and mortality in sepsis: From the HYPRESS study. *Crit. Care* **2022**, *7*, 343. [CrossRef]
31. Lyu, Q.Q.; Zheng, R.Q.; Chen, Q.H.; Yu, J.Q.; Shao, J.; Gu, X.H. Early administration of hydrocortisone, vitamin C, and thiamine in adult patients with septic shock: A randomized controlled clinical trial. *Crit. Care* **2022**, *28*, 295. [CrossRef] [PubMed]
32. Group, COVID STEROID 2 Trial; Munch, M.W.; Myatra, S.N.; Vijayaraghavan, B.K.T.; Saseedharan, S.; Benfield, T.; Wahlin, R.R.; Rasmussen, B.S.; Andreasen, A.S.; Poulsen, L.M.; et al. Effect of 12 mg vs 6 mg of Dexamethasone on the Number of Days Alive Without Life Support in Adults With COVID-19 and Severe Hypoxemia: The COVID STEROID 2 Randomized Trial. *JAMA* **2021**, *326*, 1807–1817. [PubMed]
33. Wong, H.R.; Hart, K.W.; Lindsell, C.J.; Sweeney, T.E. External Corroboration That Corticosteroids May Be Harmful to Septic Shock Endotype A Patients. *Crit. Care Med.* **2021**, *1*, e98–e101. [CrossRef] [PubMed]
34. Wong, H.R.; Cvijanovich, N.Z.; Anas, N.; Allen, G.L.; Thomas, N.J.; Bigham, M.T.; Weiss, S.L.; Fitzgerald, J.; Checchia, P.A.; Meyer, K.; et al. Developing a clinically feasible personalized medicine approach to pediatric septic shock. *Am. J. Respir. Crit. Care Med.* **2015**, *191*, 309–315. [CrossRef] [PubMed]
35. Moskowitz, A.; Huang, D.T.; Hou, P.C.; Gong, J.; Doshi, P.B.; Grossestreuer, A.V.; Andersen, L.W.; Ngo, L.; Sherwin, R.L.; Berg, K.M.; et al. ACTS Clinical Trial. Effect of Ascorbic Acid, Corticosteroids, and Thiamine on Organ Injury in Septic Shock: The ACTS Randomized Clinical Trial. *JAMA* **2020**, *324*, 642–650. [CrossRef] [PubMed]
36. Fujii, T.; Luethi, N.; Young, P.J.; Frei, D.R.; Eastwood, G.M.; French, C.J.; Deane, A.M.; Shehabi, Y.; Hajjar, L.A.; Oliveira, G.; et al. Effect of Vitamin C, Hydrocortisone, and Thiamine vs Hydrocortisone Alone on Time Alive and Free of Vasopressor Support Among Patients With Septic Shock: The VITAMINS Randomized Clinical Trial. *JAMA* **2020**, *323*, 423–431. [CrossRef] [PubMed]
37. Thompson, K.J.; Taylor, C.B.; Venkatesh, B.; Cohen, J.; Hammond, N.E.; Jan, S.; Li, Q.; Myburgh, J.; Rajbhandari, D.; Saxena, M.; et al. The ADRENAL Management Committee and Investigators and the ANZICS Clinical Trials. The cost-effectiveness of adjunctive corticosteroids for patients with septic shock. *Crit. Care Resusc.* **2020**, *22*, 191–199. [PubMed]
38. Venkatesh, B.; Finfer, S.; Cohen, J.; Rajbhandari, D.; Arabi, Y.; Bellomo, R.; Billot, L.; Glass, P.; Joyce, C.; Li, Q.; et al. Hydrocortisone Compared with Placebo in Patients with Septic Shock Satisfying the Sepsis-3 Diagnostic Criteria and APROCCHSS Study Inclusion Criteria: A Post Hoc Analysis of the ADRENAL Trial. *Anesthesiology* **2019**, *131*, 1292–1300. [CrossRef]
39. Antcliffe, D.B.; Burnham, K.L.; Al-Beidh, F.; Santhakumaran, S.; Brett, S.J.; Hinds, C.J.; Ashby, D.; Knight, J.C.; Gordon, A.C. Transcriptomic Signatures in Sepsis and a Differential Response to Steroids. From the VANISH Randomized Trial. *Am. J. Respir. Crit. Care Med.* **2019**, *99*, 980–986. [CrossRef]
40. Gordon, A.C.; Mason, A.J.; Thirunavukkarasu, N.; Perkins, G.D.; Cecconi, M.; Cepkova, M.; Pogson, D.G.; Aya, H.D.; Anjum, A.; Frazier, G.J.; et al. VANISH Investigators. Effect of early vasopressin vs norepinephrine on kidney failure in patients with septic shock: The VANISH randomized clinical trial. *JAMA* **2016**, *316*, 509. [CrossRef]

41. Coutinho, A.E.; Chapman, K.E. The anti-inflammatory and immunosuppressive effects of glucocorticoids, recent developments and mechanistic insights. *Mol. Cell Endocrinol.* **2011**, *15*, 2–13. [CrossRef] [PubMed]
42. Ritter, L.A.; Britton, N.; Heil, E.L.; Teeter, W.A.; Murthi, S.B.; Chow, J.H.; Ricotta, E.; Chertow, D.S.; Grazioli, A.; Levine, A.R. The Impact of Corticosteroids on Secondary Infection and Mortality in Critically Ill COVID-19 Patients. *J. Intensive Care Med.* **2021**, *36*, 1201–1208. [CrossRef] [PubMed]
43. Loisa, P.; Parviainen, I.; Tenhunen, J.; Hovilehto, S.; Ruokonen, E. Effect of mode of hydrocortisone administration on glycemic control in patients with septic shock: A prospective randomized trial. *Crit. Care* **2007**, *11*, R21. [CrossRef]
44. Imaizumi, T.; Nakatochi, M.; Fujita, Y.; Yamamoto, R.; Watanabe, K.; Maekawa, M.; Yamawaka, T.; Katsuno, T.; Maruyama, S. Glucocorticoid treatment is associated with ICU-acquired hypernatremia: A nested case–control study. *Clin. Exp. Nephrol.* **2021**, *5*, 131–139. [CrossRef] [PubMed]
45. Marik, P.E.; Khangoora, V.; Rivera, R.; Hooper, M.H.; Catravas, J. Hydrocortisone, Vitamin C, and Thiamine for the Treatment of Severe Sepsis and Septic Shock: A Retrospective Before-After Study. *Chest* **2017**, *151*, 1229–1238. [CrossRef]

biomedicines

MDPI

Article

ARDS Mortality Prediction Model Using Evolving Clinical Data and Chest Radiograph Analysis

Ana Cysneiros [1,2,*], Tiago Galvão [3], Nuno Domingues [3], Pedro Jorge [3], Luis Bento [1,2] and Ignacio Martin-Loeches [4]

1 Nova Medical School, Universidade de Lisboa, 1649-004 Lisbon, Portugal; luis.bento@chlc.min-saude.pt
2 Unidade de Urgência Médica, Hospital de São José, Centro Hospitalar Universitário Lisboa Central, 1169-050 Lisbon, Portugal
3 Instituto Politécnico de Lisboa/Instituto Superior de Engenharia de Lisboa, 1959-007 Lisbon, Portugal; tiagogalvao97@gmail.com (T.G.); nuno.ndomingues@isel.pt (N.D.); pedro.mendes.jorge@isel.pt (P.J.)
4 Trinity College Dublin, D02 PN40 Dublin, Ireland; drmartinloeches@gmail.com
* Correspondence: taiacysneiros@gmail.com

Abstract: Introduction: Within primary ARDS, SARS-CoV-2-associated ARDS (C-ARDS) emerged in late 2019, reaching its peak during the subsequent two years. Recent efforts in ARDS research have concentrated on phenotyping this heterogeneous syndrome to enhance comprehension of its pathophysiology. Methods and Results: A retrospective study was conducted on C-ARDS patients from April 2020 to February 2021, encompassing 110 participants with a mean age of 63.2 ± 11.92 (26–83 years). Of these, 61.2% (68) were male, and 25% (17) experienced severe ARDS, resulting in a mortality rate of 47.3% (52). Ventilation settings, arterial blood gases, and chest X-ray (CXR) were evaluated on the first day of invasive mechanical ventilation and between days two and three. CXR images were scrutinized using a convolutional neural network (CNN). A binary logistic regression model for predicting C-ARDS mortality was developed based on the most influential variables: age, PaO_2/FiO_2 ratio (P/F) on days one and three, CNN-extracted CXR features, and age. Initial performance assessment on test data (23 patients out of the 110) revealed an area under the receiver operating characteristic (ROC) curve of 0.862 with a 95% confidence interval (0.654–0.969). Conclusion: Integrating data available in all intensive care units enables the prediction of C-ARDS mortality by utilizing evolving P/F ratios and CXR. This approach can assist in tailoring treatment plans and initiating early discussions to escalate care and extracorporeal life support. Machine learning algorithms for imaging classification can uncover otherwise inaccessible patterns, potentially evolving into another form of ARDS phenotyping. The combined features of these algorithms and clinical variables demonstrate superior performance compared to either element alone.

Keywords: ARDS; imaging; machine learning; deep learning

Citation: Cysneiros, A.; Galvão, T.; Domingues, N.; Jorge, P.; Bento, L.; Martin-Loeches, I. ARDS Mortality Prediction Model Using Evolving Clinical Data and Chest Radiograph Analysis. *Biomedicines* **2024**, *12*, 439. https://doi.org/10.3390/biomedicines12020439

Academic Editor: Christian Lehmann

Received: 19 January 2024
Revised: 12 February 2024
Accepted: 14 February 2024
Published: 16 February 2024

1. Introduction

Acute Respiratory Distress Syndrome (ARDS) is a critical medical condition characterized by severe respiratory failure posing a substantial threat with high mortality rates. It can manifest in response to diverse underlying causes, including pneumonia, sepsis, trauma, or inhalation injuries [1]. ARDS is a significant concern in critical care medicine due to its potential for rapid progression and severe respiratory compromise. The Berlin criteria define ARDS by acute lung insult, bilateral chest infiltrates, and hypoxemia not fully explained by other factors [1,2].

While widely accepted, the Berlin criteria need to capture ARDS's multifaceted nature fully. This syndrome encompasses a spectrum of clinical disorders, diverse physiological abnormalities, varied radiographic presentations, multiple potential microbiological causes, and dynamic evolution over time [3,4]. ARDS exhibits a continuum from the early development of acute lung injury to meeting specific diagnostic criteria [4].

Diagnostic Criteria and Radiological Evidence

The Berlin criteria assume the use of arterial blood gases, specifically the PaO_2/FiO_2 ratio, and require radiological evidence often obtained through chest X-ray (CXR) or chest CT scans. Recently published ESICM definitions have sparked debate and alternative approaches in the medical community, challenging the reliance on chest radiography [5,6].

A prior randomized controlled trial (RCT) revealed no improvement in CXR interpretation following standardized training specific to ARDS. Alternative approaches have been debated, such as accepting unilateral opacities mandating computed tomography (CT) scans and incorporating lung ultrasound in ARDS [6].

Despite being a prevalent cause of acute respiratory failure with high morbidity and mortality, ARDS lacks proven therapeutic options beyond lung protective ventilation [7].

The underlying heterogeneity has prompted research into ARDS sub-phenotypes that may respond to specific treatments [8–10].

Recent latent class analysis (LCA) has identified sub-phenotypes with distinct clinical and biological features, emphasizing the importance of precision medicine in addressing ARDS heterogeneity [10].

The lack of a definitive diagnostic test contributes to the broad definition of ARDS, exemplified by histopathological findings in only 45% of autopsied lungs from ARDS patients [11,12].

The debate over steroids in ARDS underscores the need for phenotyping, as specific subtypes, like COVID-19-associated ARDS (C-ARDS), may benefit from steroids [13], whereas plenty of RCTs have not been able to demonstrate clear benefit from steroids in ARDS [14,15].

Prediction of ARDS remains challenging, and the heterogeneity of its etiology often hampers clinical trials. Machine learning, especially in radiology, is gaining prominence and has been employed in ARDS diagnosis [16–19] as well as studying recruitability and response to PEEP [17,20].

Our study aims to analyze evolving ventilation settings, PaO_2/FiO_2, and chest radiography in patients with C-ARDS. The hypothesis is that the combination of chest X-rays with ventilation settings and arterial blood gases can effectively predict ARDS mortality, contributing to more comprehensive understanding of this complex syndrome.

2. Materials and Methods

2.1. Study Design

A retrospective cohort study was conducted on mechanically ventilated C-ARDS patients admitted to the intensive care units of Hospital São José and Hospital Curry Cabral between April 2020 and January 2021. This is a university hospital trust with 217 beds and 55 ICU beds. Patient data were acquired through electronic medical records.

2.2. Inclusion and Exclusion Criteria

Patients on extracorporeal membrane oxygenation (ECMO) were excluded due to potential differences in ventilation data arising from lung rest ventilation strategies. The analysis focused on the first 72 h of ICU admission when SARSCOV was the sole infectious isolate. Patients diagnosed with ARDS according to the Berlin definition requiring invasive mechanical ventilation for at least 48 h were included. Exclusion criteria encompassed age below 18, pregnancy, any other contributing causes of ARDS, synchronous respiratory infection with other agents, death within the initial 48 h of ICU admission, and lack of data on ventilation settings or poor-quality chest radiographs.

2.3. Primary Outcome

The primary outcome was defined as all-cause mortality.

2.4. Data Acquisition

Patient data included age, gender, arterial blood gases, and ventilation settings during the first day (PS d1) and between 48 and 72 h (PS d3) of invasive mechanical ventilation. Portable chest radiographs (CXRs) at day one (CXR d1) and day three (CXR d3) were processed using Gaussian blur filtering and contrast-limited adaptive histogram equalization. These techniques aim to achieve noise reduction and image contrast optimization in the regions of interest and improve accuracy in image classification tasks using neural networks [21].

The lung area was segmented and properly resized, and normalized CXR d1 and CXR d3 were concatenated. A pre-trained DenseNet121 convolutional neural network (CheXNet) was employed for transfer learning, providing reliable analysis and detection of 14 thoracic-related pathologies. Deep learning features (DLFs) were extracted and coupled with clinical variables (CVs) to construct two machine learning models: logistic regression (LogReg) and a multilayer perceptron (MLP). Age, PaO_2/FiO_2 ratio on the third day (P/F d3) of invasive mechanical ventilation, and deep learning features (DLFs) were used in the final models.

2.5. Sample Split and Imputation

Sample data were randomly split into training (85 patients) and test (23 patients) groups. Missing data were imputed by the median after splitting. Imputation was individually applied to each subset of cross-validation to reduce risk of bias and variance. Other methods were not used because the dataset was relatively small, with the risk of outliers.

2.6. Statistical Analysis

Models were developed using logistic regression (LogReg), multilayer perceptron (MLP), support vector machine (SVM), and random forest (RndForest) algorithms for comparison. Sequential feature selection, employing logistic regression with five-fold cross-validation in the training group, was performed to avoid model overfitting. For comparison purposes, a model was created using only clinical data with age and P/F d3 (Model B).

3. Results

A total of 110 patients were enrolled in the study, with a mean age of 63.2 $\pm$ 11.92 (26–83 years). The gender distribution revealed an expected predominance of males, constituting 61.2% (n = 68) of the cohort. Severe ARDS was present in 25% (17) of patients. The overall mortality rate was 47.3% (52). Further epidemiological characteristics are displayed in Figure 1 and Table 1.

Figure 1 shows the majority of patients had moderate ARDS based on their P/F ratio, with severe ARDS more common in the 60–80 age group and more common in males.

Table 1 shows 61.2% of patients were male and overall mortality was 47.3%. The average age was 63.2% (min 26; max 83) and the average P/F ratio on admission was 148.5 (min 51 max 297). The average P/F ratio increased on day 3 (163.73). The average PEEP on day 1 was 12.76 and was 11.61 on day 3.

Table 2 shows how DLFs increased sensitivity from 0.643 (model B) to 0.714 (model A); specificity from 0.711 to 0.778; the positive predictive value (PPV) from 0.675 to 0.75 and the AUC from 0.701 to 0.77. Regarding the Bayesian t-test results, the probability of model A being better than model B was 0.739.

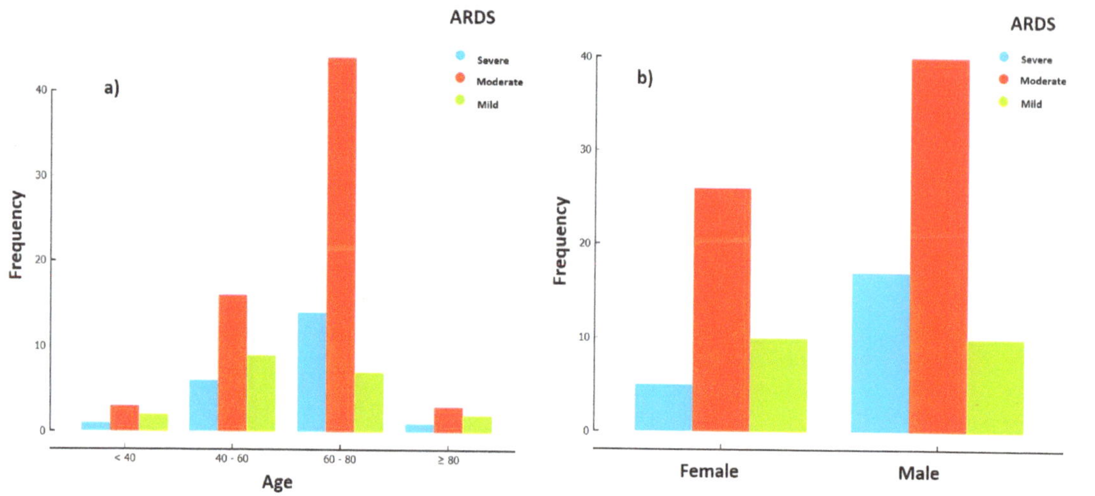

Figure 1. Distribution of ARDS severity at day 1 of IMV based on pf_d1. (**a**) Distribution of ARDS severity by age. (**b**) Distribution of ARDS severity by gender.

Table 1. Descriptive statistics of the total database.

Variables	**Categorical (n = 110)**					
Mortality (target: NS)	**Non-Survivors**		**n = 52**		**47.3%**	
	Survived		n = 58		52.7%	
Gender	Male		n = 68		61.2%	
	Female		n = 42		38.2%	
			Continuous			
	Average. ± s.d	**Median**	**Mode**	**Min–Max**	**C_v**	**Missing (%)**
Age	63.2 ±11.87	64	64	26:83	0.19	0
P/F_d1	148.50± 52.92	147.5	150	51:297	0.36	2 (2%)
P/F_d3	163.73 ± 53.07	160	118	52:365	0.32	1 (1%)
PS_d1	16.22 ± 3.19	16	16	8:26	0.20	3 (3%)
PS_d3	15.49 ± 3.50	18	16	0:26	0.23	7 (6%)
PEEP_d1	12.76 ± 3.29	14	14	6:20	0.26	0 (0%)
PEEP_d3	11.61 ± 3.49	12	12	0:22	0.30	5 (5%)
PaCO2_d1	45.84 ± 11.36	44.6	46	18.1:89	0.25	1 (1%)
PaCO2_d3	47.32 ± 9.61	46.2	40.3	26.4:90.6	0.20	2 (2%)

n = number of samples; NS = non-survivor; mean. ± d.p = mean ± standard deviation; min– max = minimum and maximum values of the variable; Cv = coefficient of variation; missing (%) = percentage of missing samples. The target class is shown in bold.

3.1. Logistic Regression Analysis

Utilizing logistic regression analysis, the most influential variables were identified as age, PaO_2/FiO_2 at day one (P/F d1), PaO_2/FiO_2 at day three (P/F d3), and deep learning image features (DLFs). Cross-validation using P/F d3, age, and DLFs demonstrated optimal performance metrics.

Table 2. Average of the performance metrics of the LogReg_A and LogReg_B models with their Bayesian *t*-test results.

Model		$AUC_{méd}$	AUC_p	CA	Ss	PPV	Sp	F1
LogReg_A		0.770	0.784	0.747	0.714	0.750	0.778	0.732
LogReg_B		0.701	0.751	0.678	0.643	0.675	0.711	0.659
Bayesian *t*-test	P_AB		0.739	0.886	0.769	0.875	0.758	0.864
	P_BA		0.261	0.114	0.231	0.122	0.242	0.136
	$P_{5\%}$_AB		0.364	0.630	0.568	0.734	0.547	0.713
	$P_{5\%}$_BA		0.062	0.026	0.104	0.204	0.107	0.056
	$P_{neg5\%}$		0.574	0.453	0.328	0.230	0.346	0.230

P_AB = probability of model A being better than B; P_BA = probability of model B being better than model A; $P_{5\%}$_AB = probability of model A being better than model B considering a negligible 5% performance difference; $P_{5\%}$_BA = probability of model B being better than A considering a negligible 5% performance difference; $P_{neg5\%}$ = probability of the models being identical considering a negligible performance difference of 5%. CA = classification accuracy; Ss = sensitivity, Sp = specificity; PPV = positive predictive value; F1 = harmonic mean of PPV and sensitivity.

3.2. Model Performance

The models incorporating DLFs exhibited an 89% probability of superior accuracy for logistic regression (LogReg) and 82% for multilayer perceptron (MLP) compared to the clinical variables-only model (Model B). Within the internal test group (23 patients), the LogReg model emerged as the most robust, yielding an area under the ROC curve (AUC) of 0.862 with a 95% confidence interval [0.654, 0.969], an accuracy of 0.783 (95% CI [0.563, 0.926]), and an F1 score of 0.783 (95% CI [0.563, 0.926]). Detailed results are shown in Table 3.

Table 3. Performance metrics of the final model LogReg_A in the training group (LogReg_train), in cross-validation (LogReg_CV), and in the test group (LogReg_Test). Δ_{CV_Train} = difference between metrics from the training group ad cross-validation. $\Delta_{Test\ Train}$ = difference between the metrics of the training group and test group.

Metrics	LogReg_Train95%CI	LogReg_CV95%CI	LogReg_Test95%CI	Δ_{CV_Train}	$\Delta_{Test\ Train}$
AUC_{med}	0.820 [0.723–0.894]	0.770 [0.667–0.853]	0.862 [0.654–0.969]	−0.092	0.042
AUC_p	0.820	0.784	0.862	−0.036	0.042
CA	0.782 [0.681–0.863]	0.747 [0.643–0.834]	0.783 [0.563–0.926]	−0.035	−0.0
F1	0.776 [0.672–0.859]	0.732 [0.623–0.824]	0.783 [0.563–0.926]	−0.044	0.007
PPV	0.767 [0.623–0889]	0.750 [0.588–0.873]	0.692 [0.385–0.909]	−0.026	0.084
NPV	0.795 [0.647- 0.902]	0.744 [0.596- 0.861]	0.900 [0.555–0.997]	−0.051	0.105
Ss	0.787 [0.633–0.898]	0.714 [0.554–0.843]	0.900 [0.555–0.997]	−0.073	0.113
Sp	0.778 [0.629–0.888]	0.778 [0.629–0.888]	0.692 [0.385–0.909]	0.000	−0.086

positive predictive value (PPV), negative predictive value (NPV), sensitivity (Ss), specificity (Sp), F1 = harmonic mean of PPV, sensitivity, and classification accuracy (CA).

3.3. Comparative Analysis

The same analysis was repeated using only clinical variables and P/F d3, revealing an AUC of 0.77. Notably, enhanced classification performance was observed when incorporating CNN-extracted image features.

3.4. Cross-Validation Significance

Cross-validation, which is crucial for reducing bias and accounting for outliers within the same population, was employed. This approach is particularly significant in small-population studies, ensuring robustness in the analysis and interpretation of the results, as shown in Figure 2 and Table 4.

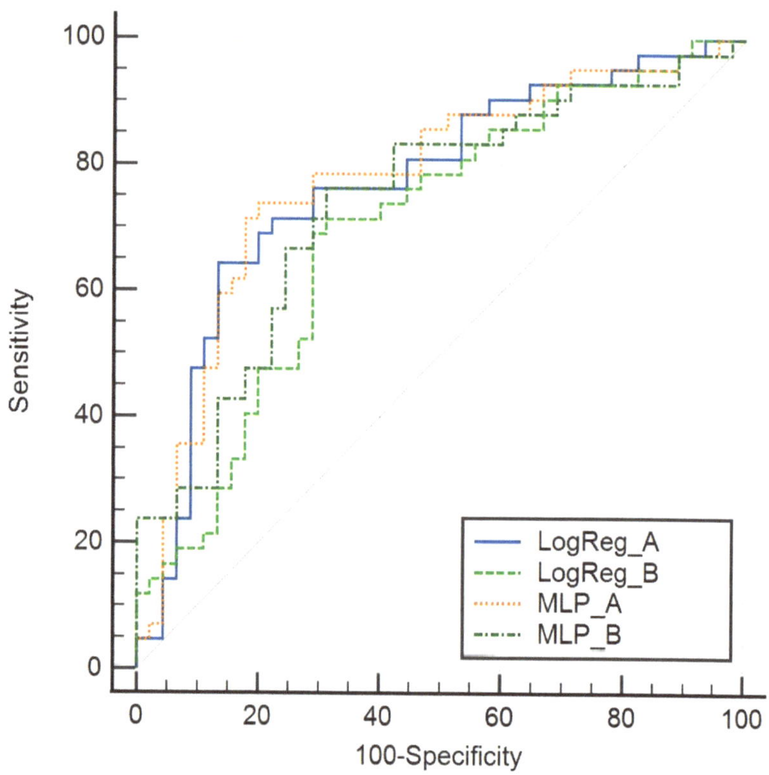

Figure 2. Cross-validation sensitivity and specificity of model A and model B by logistic regression and MLP.

Figure 2 shows the cross-validation sensitivity and specificity of both models by logistic regression and MLP, with model A, which includes DLF, performing better than model B.

Table 4 uses cross-validation, which is important when using small populations. Again, the contribution of DLF is sustained, with AUC improving from 0.701 (CI 0.593–0.794) to 0.77 (CI 0.667–0.853). Improvement was greater in PPV, which rose to 0.75 (CI 0.558–0.839) from 0.675 (CI 0.509–0.814), and than NPV, which improved from 0.714 (CI 0.554–0.843) to 0.744 (CI 0.596–0.861).

Figure 3 shows the individual contribution of the three selected features to the final model using logistic regression.

Table 4. Cross-validation analysis by logistic regression.

Metric	LogReg_A	LogReg_B	Δ_{A-B}
AUC_{med}	0.770 95%CI [0.667–0.853]	0.701 95%CI [0.593–0.794]	0.069
CA	0.747 95%CI [0.643–0.834]	0.678 95%CI [0.569–0.774]	0.069
PPV	0.750 95%CI [0.588–0.873]	0.675 95%CI [0.509–0.814]	0.075
Sensitivity	0.714 95%CI [0.554–0.843]	0.643 95%CI [0.480–0.784]	0.071
NPV	0.744 95%CI [0.596–0.861]	0.714 95%CI [0.554- 0.843]	0.081
Specificity	0.778 95%CI [0.629–0.888]	0.711 95%CI [0.560–0.834]	0.067
F1	0.732 95%CI [0.623–0.824]	0.659 95%CI [0.546–0.760]	0.073

positive predictive value (PPV); negative predicted value (NPV); sensitivity (Ss); specificity (Sp); classification accuracy (CA); F1 = harmonic mean of PPV and sensitivity.

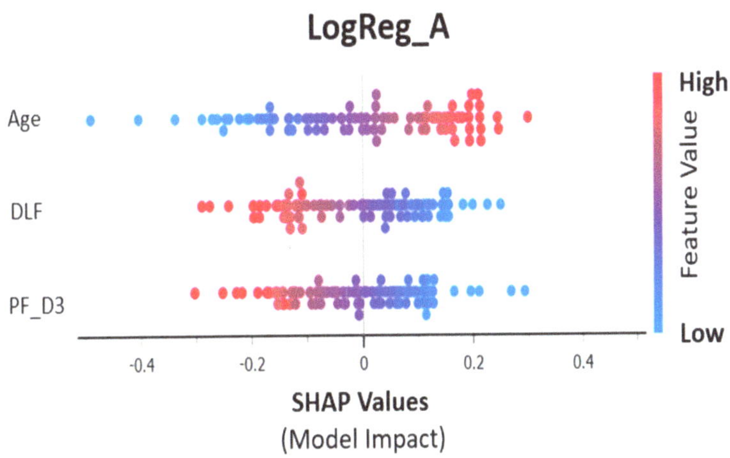

Figure 3. SHAP values (contribution to training prediction).

Figure 3 shows the contribution of the three selected features to the model using logistic regression. Increasing age and decreasing P/F are therefore associated with mortality (the positive outcome). DLFs result from the concatenated images of CXR at day 1 and day 3.

4. Discussion

In the evolving landscape of Acute Respiratory Distress Syndrome (ARDS) research, the utilization of machine learning techniques, particularly in conjunction with chest radiography, holds promise for providing valuable insights into the prediction of mortality and severity in patients with COVID-19-associated ARDS (C-ARDS). This study explored the association between image features extracted from chest radiographs and patient outcomes, emphasizing the evolving nature of chest imaging in the first 72 h of invasive mechanical ventilation. Our main message is that we proved that integrating deep learning image features in the logistic regression model exhibited superior predictive accuracy, providing valuable insights into mortality prediction in C-ARDS patients. The robust performance metrics, especially within the internal test group, underscore the potential clinical utility of the proposed model.

In the dynamic and continually advancing field of ARDS research, integrating machine learning techniques, particularly in tandem with chest radiography, presents a transformative avenue with substantial promise. This synergistic approach has the potential to yield profound insights into predicting not only the mortality outcomes but also the severity of ARDS in afflicted patients. The confluence of machine learning and chest radiography stands out as a cutting-edge paradigm poised to enhance our understanding of ARDS's intricate dynamics and nuanced manifestations, ultimately contributing to more effective and personalized approaches to patient management.

The evolving landscape of ARDS research reflects a growing recognition of the complexities inherent in this critical medical condition. Traditional methodologies have encountered challenges in addressing the heterogeneity of ARDS, emphasizing the need for innovative and sophisticated techniques to unravel its multifaceted nature. In this context, the amalgamation of machine learning and chest radiography emerges as a revolutionary strategy, offering a comprehensive and nuanced perspective on the predictive factors influencing both mortality and severity in ARDS patients.

Machine learning, with its capacity to discern patterns and relationships within vast datasets, complements the intricacies of ARDS by providing a data-driven framework for analysis. Integrating these advanced computational methods with chest radiography, a widely accessible imaging modality, establishes a powerful synergy. This combined approach capitalizes on the detailed information embedded in radiographic images, enabling the identification of subtle yet clinically significant features that may elude conventional diagnostic and prognostic assessments.

The promise lies not only in the ability to predict mortality outcomes but also in gauging the severity of ARDS, a crucial aspect that influences the trajectory of patient care. By harnessing the potential of machine learning algorithms to analyze intricate radiographic details, the predictive model becomes more adept at discerning the nuances of disease progression, thereby contributing to more nuanced understanding of ARDS severity.

This integrated approach is not merely confined to a technological juxtaposition but represents a fundamental shift in the paradigm of ARDS research. It transcends the conventional boundaries of diagnostic and prognostic methodologies, offering a holistic and data-driven framework that aligns with the evolving complexities of ARDS pathophysiology. As a result, this innovative synthesis of machine learning and chest radiography stands poised to redefine the landscape of ARDS research, ushering in a new era of precision medicine and personalized patient care.

The challenges in ARDS treatment have been underscored by its inherent heterogeneity. The RECOVERY trial [13] demonstrated a mortality benefit of steroids in mechanically ventilated patients fulfilling the Berlin criteria, particularly in those with COVID-19-associated ARDS (C-ARDS), suggesting a potential subgroup homogeneity. However, the applicability of steroids across all ARDS cases remains uncertain, as evidenced by varying outcomes in ARDS secondary to influenza [22,23].

Our study delves into the predictive capacity of deep learning features extracted from chest radiographs, surpassing the predictive capability of the P/F ratio in terms of mortality. This finding aligns with the evolving trend in ARDS research, which emphasizes the importance of integrating lung morphology assessments in patient management. Studies have traditionally focused on radiographic assessment of lung edema (RALE), which was validated using patients in the ARDS Network Fluid and Catheter Treatment Trial [24]. Studies have found the RALE score to be correlated with ARDS severity [25,26] and survival [27], but this score requires specific training to reduce observer variability.

Recent studies using imaging patterns to tailor ventilation strategies have had mixed outcomes. The LIVE trial, for instance, indicated that personalization based on CT-based image classification did not decrease mortality, potentially due to misclassification and subsequent mismatch in ventilator strategies [28]. Automated approaches, as performed in our study, identify regions of interest with less risk of misclassification.

Our study sheds light on the evolving nature of chest imaging within the first 72 h of invasive mechanical ventilation, revealing a strong correlation with mortality. Unlike baseline images, [29,30], this temporal relationship aligns with recent studies highlighting the prognostic value of changes in imaging parameters over time [29].

Lung ultrasound has also emerged as promising in distinguishing between focal and non-focal ARDS [31,32], and trials are ongoing regarding lung ultrasound patterns and personalized mechanical ventilation. However, lung ultrasound is operator-dependent and more time-consuming.

Integrating multi-source data becomes imperative as ARDS research transitions towards a phenotyping strategy. This study, leveraging data readily available in the ICU, demonstrates our potential to achieve a robust predictive model. Nevertheless, the study's limitations, including its retrospective nature and relatively small population size, warrant prospective validation to enhance its clinical utility; an external validation cohort is underway, which will include ARDS patients from a Dublin ICU. The data assessed will include more outcomes such as ventilator-free days and length of ICU stay.

Given its high volume of images and cost-effectiveness, chest radiography emerges as an appealing modality for machine learning applications. Unlike chest tomography, bedside chest radiographs offer the advantage of being readily accessible and conducive to repeated examinations, facilitating the assessment of disease progression. However, limitations in available data necessitated the selection of clinical variables. Notably, comprehensive data on ventilation parameters, including driving pressure and plateau pressure, were not uniformly obtainable within the specified timeframe. Consequently, the PaO_2/FiO_2 ratio (P/F ratio) was chosen, despite its susceptibility to influence from positive end-expiratory pressure (PEEP).

5. Conclusions

In conclusion, our findings underscore the significance of integrating chest radiography and machine learning in early mortality prediction in C-ARDS. The evolving aspect of chest imaging, particularly within the first 72 h of invasive mechanical ventilation, emerges as a critical determinant of patient outcomes. As ARDS research advances towards a phenotyping strategy, we anticipate future studies will build upon our findings, combining multi-source data to refine treatment strategies for specific patient subgroups. Prospective validation is essential for establishing the clinical applicability of our predictive model, ultimately enabling early discussions and strategic planning for patients at higher risk of mortality in the critical care setting.

Author Contributions: Conceptualization, A.C.; Methodology, T.G.; Formal analysis, I.M.-L.; Investigation, A.C., T.G., N.D., P.J. and I.M.-L.; Resources, N.D. and P.J.; Data curation, T.G.; Writing—original draft, A.C.; Supervision, L.B. and I.M.-L. All authors have read and agreed to the published version of the manuscript.

Funding: This research was partially funded by Predictive Models of COVID-19 Outcomes for Higher Risk Patients Towards Precision Medicine (PREMO) project, by Fundação para a Ciência e Tecnologia (FCT) grant DSAIPA/DS/0117/2020.

Institutional Review Board Statement: The study was conducted in accordance with the Declaration of Helsinki, and approved by the Ethics Committee of Comissão de Ética para a Saúde 1043/2021 20th of May 2020 for studies involving humans.

Informed Consent Statement: Individual patient consent was waived considering no intervention was conducted and data was fully anonymised and cannot be identified.

Data Availability Statement: The debase is still being used for research and other outcomes but will be shared in the future.

Conflicts of Interest: The authors declare no conflict of interest.

References

1. Force, A.D.; Ranieri, V.M.; Rubenfeld, G.D.; Thompson, B.T.; Ferguson, N.D.; Caldwell, E.; Fan, E.; Camporota, L.; Slutsky, A.S. Acute respiratory distress syndrome: The Berlin Definition. *JAMA* **2012**, *307*, 2526–2533. [CrossRef]
2. Matthay, M.A.; Zemans, R.L.; Zimmerman, G.A.; Arabi, Y.M.; Beitler, J.R.; Mercat, A.; Herridge, M.; Randolph, A.G.; Calfee, C.S. Acute respiratory distress syndrome. *Nat. Rev. Dis. Primers* **2018**, *5*, 18. [CrossRef] [PubMed]
3. Ware, L.B.; Matthay, M.A. The acute respiratory distress syndrome. *N. Engl. J. Med.* **2000**, *342*, 1334–1349. [CrossRef] [PubMed]
4. Meyer, N.J.; Gattinoni, L.; Calfee, C.S. Acute respiratory distress syndrome. *Lancet* **2021**, *398*, 622–637. [CrossRef]
5. Grasselli, G.; Calfee, C.S.; Camporota, L.; Poole, D.; Amato, M.B.P.; Antonelli, M.; Arabi, Y.M.; Baroncelli, F.; Beitler, J.R.; Bellani, G.; et al. ESICM guidelines on acute respiratory distress syndrome: Definition, phenotyping and respiratory support strategies. *Intensiv. Care Med.* **2023**, *49*, 727–759. [CrossRef] [PubMed]
6. Matthay, M.A.; Arabi, Y.; Arroliga, A.C.; Bernard, G.; Bersten, A.D.; Brochard, L.J.; Calfee, C.S.; Combes, A.; Daniel, B.M.; Ferguson, N.D.; et al. A New Global Definition of Acute Respiratory Distress Syndrome. *Am. J. Respir. Crit. Care Med.* **2023**, *209*, 37–47. [CrossRef]
7. Yndrome, S.; Etwork, N. Ventilation with lowertidal volumesas compared with traditionaltidal volumes for acute lung injury and the acute respiratory distress syndrome. *N. Engl. J. Med.* **2000**, *342*, 1301–1308.
8. Bos, L.D.J.; Ware, L.B. Acute respiratory distress syndrome: Causes, pathophysiology, and phenotypes. *Lancet* **2022**, *400*, 1145–1156. [CrossRef]
9. Wildi, K.; Livingstone, S.; Palmieri, C.; LiBassi, G.; Suen, J.; Fraser, J. The discovery of biological subphenotypes in ARDS: A novel approach to targeted medicine? *J. Intensiv. Care* **2021**, *9*, 14. [CrossRef]
10. Maddali, M.V.; Churpek, M.; Pham, T.; Rezoagli, E.; Zhuo, H.; Zhao, W.; He, J.; Delucchi, K.L.; Wang, C.; Wickersham, N.; et al. Validation and utility of ARDS subphenotypes identified by machine-learning models using clinical data: An observational, multicohort, retrospective analysis. *Lancet Respir. Med.* **2022**, *10*, 367–377. [CrossRef]
11. Thille, A.W.; Esteban, A.; Fernández-Segoviano, P.; Rodriguez, J.-M.; Aramburu, J.-A.; Peñuelas, O.; Cortés-Puch, I.; Cardinal-Fernández, P.; Lorente, J.A.; Frutos-Vivar, F. Comparison of the Berlin Definition for Acute Respiratory Distress Syndrome with Autopsy. *Am. J. Respir. Crit. Care Med.* **2013**, *187*, 761–767. [CrossRef] [PubMed]
12. Reilly, J.; Calfee, C.; Christie, J. Acute Respiratory Distress Syndrome Phenotypes. *Semin. Respir. Crit. Care Med.* **2019**, *40*, 19–30. [CrossRef] [PubMed]
13. The Recovery Collaborative Group. Dexamethasone in Hospitalized Patients with Covid-19. *N. Engl. J. Med.* **2021**, *384*, 693–704. [CrossRef]
14. Lewis, S.R.; Pritchard, M.W.; Thomas, C.M.; Smith, A.F. Pharmacological agents for adults with acute respiratory distress syndrome. *Cochrane Database Syst. Rev.* **2019**, *7*, CD004477. [CrossRef] [PubMed]
15. Villar, J.; Ferrando, C.; Martínez, D.; Ambrós, A.; Muñoz, T.; Soler, J.A.; Aguilar, G.; Alba, F.; González-Higueras, E.; Conesa, L.A.; et al. Dexamethasone treatment for the acute respiratory distress syndrome: A multicentre, randomised controlled trial. *Lancet Respir. Med.* **2020**, *8*, 267–276. [CrossRef] [PubMed]
16. Le, S.; Pellegrini, E.; Green-Saxena, A.; Summers, C.; Hoffman, J.; Calvert, J.; Das, R. Supervised machine learning for the early prediction of acute respiratory distress syndrome (ARDS). *J. Crit. Care* **2020**, *60*, 96–102. [CrossRef] [PubMed]
17. Bitker, L.; Talmor, D.; Richard, J.C. Imaging the acute respiratory distress syndrome: Past, present and future. *Intensiv. Care Med.* **2022**, *48*, 995–1008. [CrossRef] [PubMed]
18. Reamaroon, N.; Sjoding, M.W.; Gryak, J.; Athey, B.D.; Najarian, K.; Derksen, H. Automated detection of acute respiratory distress syndrome from chest X-Rays using Directionality Measure and deep learning features. *Comput. Biol. Med.* **2021**, *134*, 104463. [CrossRef]
19. Sjoding, M.W.; Taylor, D.; Motyka, J.; Lee, E.; Co, I.; Claar, D.; I McSparron, J.; Ansari, S.; Kerlin, M.P.; Reilly, J.P.; et al. Deep learning to detect acute respiratory distress syndrome on chest radiographs: A retrospective study with external validation. *Lancet Digit. Healt.* **2021**, 3, e340–e348. [CrossRef]
20. Filippini, D.F.L.; Di Gennaro, E.; van Amstel, R.B.E.; Beenen, L.F.M.; Grasso, S.; Pisani, L.; Bos, L.D.J.; Smit, M.R. Latent class analysis of imaging and clinical respiratory parameters from patients with COVID-19-related ARDS identifies recruitment subphenotypes. *Crit. Care* **2022**, *26*, 1–10. [CrossRef]
21. Giełczyk, A.; Marciniak, A.; Tarczewska, M.; Lutowski, Z. Pre-processing methods in chest X-ray image classification. *PLoS ONE* **2022**, *17*, e0265949. [CrossRef]
22. Moreno, G.; on behalf of the GETGAG Study Group; Rodríguez, A.; Reyes, L.F.; Gomez, J.; Sole-Violan, J.; Díaz, E.; Bodí, M.; Trefler, S.; Guardiola, J.; et al. Corticosteroid treatment in critically ill patients with severe influenza pneumonia: A propensity score matching study. *Intensiv. Care Med.* **2018**, *44*, 1470–1482. [CrossRef]
23. Tsai, M.-J.; For Taiwan Severe Influenza Research Consortium (TSIRC) Investigators; Yang, K.-Y.; Chan, M.-C.; Kao, K.-C.; Wang, H.-C.; Perng, W.-C.; Wu, C.-L.; Liang, S.-J.; Fang, W.-F.; et al. Impact of corticosteroid treatment on clinical outcomes of influenza-associated ARDS: A nationwide multicenter study. *Ann. Intensiv. Care* **2020**, *10*, 26. [CrossRef]
24. National Heart, Lung, and Blood Institute Acute Respiratory Distress Syndrome (ARDS) Clinical Trials Network; Wiedemann, H.P.; Wheeler, A.P.; Bernard, G.R.; Thompson, B.T.; Hayden, D.; Deboisblanc, B.; Connors, A.F.J.; Hite, R.D.; Harabin, A.L. Comparison of Two Fluid-Management Strategies in Acute Lung Injury. *N. Engl. J. Med.* **2006**, *354*, 2564–2575. [CrossRef]

25. Warren, M.A.; Zhao, Z.; Koyama, T.; Bastarache, J.A.; Shaver, C.M.; Semler, M.W.; Rice, T.W.; Matthay, M.A.; Calfee, C.S.; Ware, L.B. Severity scoring of lung oedema on the chest radiograph is associated with clinical outcomes in ARDS. *Thorax* **2018**, *73*, 840–846. [CrossRef]
26. Sedhai, Y.R.; Yuan, M.; Ketcham, S.W.; Co, I.; Claar, D.D.; McSparron, J.I.; Prescott, H.C.; Sjoding, M.W. Validating Measures of Disease Severity in Acute Respiratory Distress Syndrome. *Ann. Am. Thorac. Soc.* **2021**, *18*, 1211–1218. [CrossRef]
27. Jabaudon, M.; Audard, J.; Pereira, B.; Jaber, S.; Lefrant, J.-Y.; Blondonnet, R.; Godet, T.; Futier, E.; Lambert, C.; Bazin, J.-E.; et al. Early Changes Over Time in the Radiographic Assessment of Lung Edema Score Are Associated With Survival in ARDS. *Chest* **2020**, *158*, 2394–2403. [CrossRef] [PubMed]
28. Constantin, J.-M.; Jabaudon, M.; Lefrant, J.-Y.; Jaber, S.; Quenot, J.-P.; Langeron, O.; Ferrandière, M.; Grelon, F.; Seguin, P.; Ichai, C.; et al. Personalised mechanical ventilation tailored to lung morphology versus low positive end-expiratory pressure for patients with acute respiratory distress syndrome in France (the LIVE study): A multicentre, single-blind, randomised controlled trial. *Lancet Respir. Med.* **2019**, *7*, 870–880. [CrossRef] [PubMed]
29. Valk, C.M.A.; Zimatore, C.; Mazzinari, G.; Pierrakos, C.; Sivakorn, C.; Dechsanga, J.; Grasso, S.; Beenen, L.; Bos, L.D.J.; Paulus, F.; et al. The Prognostic Capacity of the Radiographic Assessment for Lung Edema Score in Patients With COVID-19 Acute Respiratory Distress Syndrome—An International Multicenter Observational Study. *Front. Med.* **2022**, *8*, 772056. [CrossRef] [PubMed]
30. Herrmann, J.; Adam, E.H.; Notz, Q.; Helmer, P.; Sonntagbauer, M.; Ungemach-Papenberg, P.; Sanns, A.; Zausig, Y.; Steinfeldt, T.; Torje, I.; et al. COVID-19 Induced Acute Respiratory Distress Syndrome—A Multicenter Observational Study. *Front. Med.* **2020**, *7*, 599533. [CrossRef] [PubMed]
31. Mongodi, S.; Santangelo, E.; Bouhemad, B.; Vaschetto, R.; Mojoli, F. Personalised mechanical ventilation in acute respiratory distress syndrome: The right idea with the wrong tools? *Lancet Respir. Med.* **2019**, *7*, e38. [CrossRef] [PubMed]
32. Pierrakos, C.; Smit, M.R.; Pisani, L.; Paulus, F.; Schultz, M.J.; Constantin, J.-M.; Chiumello, D.; Mojoli, F.; Mongodi, S.; Bos, L.D.J. Lung Ultrasound Assessment of Focal and Non-focal Lung Morphology in Patients With Acute Respiratory Distress Syndrome. *Front. Physiol.* **2021**, *12*, 1482. [CrossRef] [PubMed]

biomedicines

MDPI

Article

Point-of-Care Serum Proenkephalin as an Early Predictor of Mortality in Patients Presenting to the Emergency Department with Septic Shock

Christos Verras [1,*], Sofia Bezati [1], Vasiliki Bistola [2], Ioannis Ventoulis [3], Dionysis Matsiras [1], Sotirios Tsiodras [4], John Parissis [1] and Effie Polyzogopoulou [1]

1 University Emergency Department, Attikon University Hospital, 12462 Athens, Greece; sofiabezati@gmail.com (S.B.); mats.dionysis@gmail.com (D.M.); jparissis@yahoo.com (J.P.); effiepol@med.uoa.gr (E.P.)
2 2nd Cardiology Department, Attikon University Hospital, 12462 Athens, Greece; vasobistola@yahoo.com
3 Department of Occupational Therapy, University of Western Macedonia, 50200 Ptolemaida, Greece; iventoulis@uowm.gr
4 4th Department of Internal Medicine, Attikon University Hospital, 12462 Athens, Greece; sotirios.tsiodras@gmail.com
* Correspondence: christos.verras@gmail.com; Tel.: +30-6936585677

Abstract: Background: The aim of the present study is to investigate the prognostic utility of point-of-care (POC)-measured proenkephalin (PENK), a novel biomarker, in terms of predicting in-hospital mortality in patients presenting to the emergency department (ED) with septic shock. Methods: Bedside PENK was measured in consecutive patients presenting to the ED with septic shock according to the Sepsis-3 clinical criteria. The association of PENK with inflammatory and routine biomarkers, and its role as a predictor of in-hospital mortality, was examined. Results: Sixty-one patients with septic shock [53% females, median age 83 years (IQR 71–88)] were evaluated. Median (IQR) values of creatinine, plasma lactate, soluble urokinase plasminogen activator receptor (SuPAR), procalcitonin and PENK were 1.7 (1.0–2.9) mg/dL, 3.6 (2.1–6.8) mmol/L, 13.1 (10.0–21.4) ng/mL, 2.06 (0.84–3.49) ng/mL, and 205 (129–425) pmol/L, respectively. LogPENK significantly correlated with LogLactate (rho = 0.369, $p = 0.004$), LogCreatinine (rho = 0.537, $p < 0.001$), LogProcalcitonin (rho = 0.557, $p < 0.001$), and LogSuPAR (rho = 0.327, $p = 0.011$). During hospitalization, 39/61 (64%) patients died. In a multivariable logistic regression model, logPENK was an independent predictor of in-hospital mortality (OR 11.9, 95% CI: 1.7–84.6, $p = 0.013$). Conclusion: POC PENK levels measured upon presentation to the ED strongly correlated with metabolic, renal and inflammatory biomarkers, and may serve as a predictor of in-hospital mortality in patients with septic shock.

Keywords: septic shock; point-of-care; biomarker; proenkephalin; emergency department

Citation: Verras, C.; Bezati, S.; Bistola, V.; Ventoulis, I.; Matsiras, D.; Tsiodras, S.; Parissis, J.; Polyzogopoulou, E. Point-of-Care Serum Proenkephalin as an Early Predictor of Mortality in Patients Presenting to the Emergency Department with Septic Shock. *Biomedicines* **2024**, *12*, 1004. https://doi.org/10.3390/biomedicines12051004

Academic Editor: Carlo Santambrogio

Received: 30 March 2024
Revised: 21 April 2024
Accepted: 30 April 2024
Published: 2 May 2024

1. Introduction

Sepsis and septic shock are critical conditions with high morbidity and mortality rates, and pose a significant burden on healthcare systems worldwide, thus necessitating prompt and effective management strategies [1]. According to the Third International Consensus definition, "sepsis is a life threatening clinical syndrome characterized by organ dysfunction induced by dysregulated host response to infection". Septic shock refers to the subgroup of patients with sepsis accompanied by severe cardiovascular, cellular and metabolic compromise, as evidenced by the need for vasopressors to achieve a mean arterial pressure (MAP) ≥65 mmHg and by hyperlactemia (lactate > 2 mmol/L) [1]. In the most recent analysis for the Global Burden of Disease that evaluated data from 195 countries, the global incidence of sepsis was estimated at 48.9 million (95% uncertainty interval [UI] 38.9–62.9) cases of sepsis and 11.0 million (10.1–12.0) sepsis-related deaths in 2017, with diarrheic and lower respiratory infections being the leading causes [2]. Regarding hospital-treated sepsis, a recent meta-analysis reported worldwide estimates of sepsis and severe sepsis up to 31.5 million and 19.4 million cases, respectively, leading to 5.3 million deaths annually [3]. In another study published by the Intensive Care over Nations (ICON) audit, the most prevalent sources of sepsis were respiratory tract and abdominal infections. Notably, 51.6% of patients had more than one microorganism identified, and most common infectious organisms were Gram-negative (67.1%) and Gram-positive (49.8%) bacteria, while fungi and viruses represented less prevalent infectious agents with a prevalence of 12.9% and 2.9%, respectively [4]. Although sepsis had been mostly attributed to Gram-negative bacteria in the past, recent data point towards a rise of Gram-positive bacterial, fungal and viral infections [5,6].

Inflammatory response in the context of sepsis is governed by several pathophysiological mechanisms that lead to multi-organ dysfunction and death. Excessive cytokine production and complement activation result in tissue hypoxia, cytopathic injury, mitochondrial dysfunction, coagulopathy and derangements in the vascular endothelium [7,8]. The Sequential Organ Failure Assessment (SOFA) scoring system is mostly used to identify the severity of organ insult considering certain respiratory, circulatory, coagulation, hepatic, neurological and renal parameters [9]. Risk factors for the development of sepsis include older age, higher Simplified Acute Physiology II Score (SAPS II), cancer, advanced chronic heart failure, cirrhosis, the need for invasive ventilation or renal replacement therapy, and infection with *Acinetobacter* spp. [4]. Furthermore, older age, high procalcitonin and lactate levels, higher SOFA and Acute Physiology And Chronic Health Evaluation-II (APACHE-II) scores, lactate clearance rate, combined fungal infection and Gram-negative bacterial infection, acute kidney injury (AKI), development of acute respiratory distress syndrome, and disseminated intravascular coagulation, as well as patient comorbidities and immunosuppression, have been identified as prognostic factors for poor outcomes [10,11]. Early identification of septic patients who are at risk of poor outcomes is crucial in order to optimize care of individuals presenting to the emergency department (ED) with sepsis or septic shock. AKI is one of the most frequent complications in septic patients while, at the same time, sepsis represents one of the leading causes of AKI in critically ill patients. Sepsis-induced AKI is associated with high mortality and increased healthcare costs [12,13].

Recently, biomarkers have emerged as useful tools for risk stratification, prognostication, and guidance of therapeutic decisions in septic patients. Among numerous biomarkers studied in this context, serum proenkephalin (PENK) has received attention for its potential prognostic value in predicting clinical outcomes in critically ill patients. PENK is a member of the enkephalin family, a group of endogenous opioid peptides which are expressed in a variety of tissues and act by binding mainly to δ (delta)-opioid receptors, thereby activating them. Enkephalins are involved in various physiological processes, such as modulation of pain perception, cellular growth, preconditioning, immune function and inflammation [14,15]. PENK, also termed proenkephalin A 119–159, is a 41 amino acid-long peptide derived as a by-product fragment from the cleavage of a precursor peptide called preproenkephalin A. Apart from PENK, cleavage of this precursor molecule yields several biologically active enkephalins, with leucine-enkephalin (Leu-Enk) and methionine-enkephalin (Met-Enk) being the two main functionally mature forms [15–17]. Given the fact that biologically active enkephalins are extremely unstable molecules and difficult to measure in plasma due to their short half-life, PENK has emerged as a surrogate marker of its biologically active counterparts owing to its in vitro stability and long half-life. Elevated levels of PENK have been associated with various acute and chronic medical conditions, including heart failure, AKI, and sepsis [17–19].

The aim of the present study was to investigate the association of point-of-care (POC)-measured PENK with metabolic, inflammatory and renal biomarkers, as well as its prognostic utility in terms of predicting in-hospital mortality in patients presenting to the ED with septic shock.

2. Materials and Methods

2.1. Population

This prospective observational study was conducted at the ED of Attikon University Hospital, Athens, Greece, a tertiary university hospital that provides all clinical services. Consecutive patients presenting with septic shock to the ED, between May 2022 and May 2023, were enrolled in the study. Septic shock was defined according to the Sepsis-3 clinical criteria proposed in the third international consensus definitions for sepsis and septic shock [1,20] (Supplementary Materials, Table S1). Exclusion criteria were: age less than 18 years, hemorrhagic, refractory or anaphylactic shock, refusal to participate in the study, and chronic administration of corticosteroids. All patients or their next of kin provided written informed consent before enrolment in the study. The study was performed in accordance with the Declaration of Helsinki, and was approved by the institutional review board (IRB) of the Attikon University Hospital (IRB number: EBΔ282/10-05-2022).

2.2. Study Design

The following data were obtained from each participating patient: demographics including age and sex; comorbidities; vital signs including respiratory rate (RR), pulse oximetry blood oxygen saturation (SpO2), systolic blood pressure (SBP), diastolic blood pressure (DPB), heart rate (HR) and body temperature; a 12-lead electrocardiogram (ECG); level of consciousness using the Glasgow Coma Scale [21]; laboratory assessments encompassing blood gas analysis, complete blood count, and biochemistry which included urea, creatinine, glucose, serum bilirubin, aspartate aminotransferase (AST), alanine aminotransferase (ALT), high sensitivity troponin T (hs-TnT), C-reactive protein (CRP), procalcitonin, soluble urokinase plasminogen activator receptor (SuPAR), and fibrinogen; estimated glomerular filtration rate (eGFR) using the Modification of Diet in Renal Disease (MDRD) formula [22]; National Early Warning Score (NEWS) [23]; and SOFA score [9].

All routine laboratory testing was performed at the hospital's central laboratory. Furthermore, a whole blood sample was collected for the measurement of PENK A 119–159 (PenKid), at the patient's bedside upon arrival at the ED resuscitation room, in the first 30 min upon presentation, using a commercially available point-of-care device (POC), the Nexus IB10 POC (Sphingotec GmbH, Hennigsdorf, Germany), which provides test results via whole blood in 20 min. The Nexus IB10 is an in vitro immunology analyzer that has a built-in centrifugation capability to quantitatively measure analytes in whole blood or plasma samples on a consumable disc [24]. Sphingotest®penKid®is a non-automated immunoluminometric assay (ILMA) for the in vitro diagnostic quantitative measurement of PENK A 119–159 in human ethylenediamine tetraacetic acid (EDTA) plasma [25].

After initial management at the ED, patients were admitted either to the intensive care unit (ICU) or to a high dependency unit (HDU), and were followed throughout their length of in-hospital stay, up to the time point of either hospital discharge or death.

2.3. Statistical Analysis

Statistical analysis was performed using SPSS version 29.0 (SPSS, Inc., Chicago, IL, USA). Categorical variables are summarized as frequencies and percentages, and quantitative variables are presented as median with interquartile range (IQR). Differences of qualitative parameters between patient subgroups of survivors and non-survivors were tested by using chi-square test, while differences of quantitative parameters were assessed with the use of Mann–Whitney test. Pearson's or Spearman's correlation coefficients were used to investigate correlations between normally or non-normally distributed continuous variables, respectively. Due to the highly skewed distribution of values of laboratory variables, as tested by the Kolmogorov–Smirnov test, they were log-transformed for simple correlations and regression analyses. Outcomes examined were the incidence of AKI, defined as an increase in creatinine of $\geq$0.3 mg/dl or $\geq$50% within 48 h relative to the initial value, according to the Kidney Disease Improving Global Outcomes (KDIGO) criteria [26] (Supplementary Materials, Table S2), and in-hospital mortality. Receiver operator characteristics (ROC) analysis was used to assess the ability of PENK to predict outcomes. A multivariable logistic regression model with a backward stepwise algorithm was used to investigate independent predictors of in-hospital mortality. A p value of <0.05 was considered statistically significant.

3. Results

3.1. Patient Characteristics

Between May 2022 and May 2023, 61 consecutive patients with septic shock were prospectively enrolled upon presentation to the ED. Patient characteristics are shown in Tables 1 and 2. Median PENK (IQR) was 205 (129–425) pmol/L. Median Glasgow Coma Scale was 12 (9–15), NEWS Score 12 (10–14), and SOFA Score 6 (4–9). The final diagnoses were: lower respiratory tract infection in 44 (72.1%) patients, urinary tract infection in 10 (16.1%) patients, acute cholecystitis in 2 (3.3%) patients, intra-abdominal infection in 3 (4.9%) patients and of unknown cause in 2 (3.3%) patients.

Table 1. Clinical characteristics of study patients.

	All Patients (n = 61)	Survivors (n = 22)	Non-Survivors (n = 39)	p-Value
Age, years	83 (71–88)	77 (66–87)	84 (76–88)	0.103
Sex (female), n (%)	32 (53)	9 (41)	23 (59)	0.175
Vital signs				
Temperature, °C	37.8 (36.7–38.3)	37.8 (36.4–38.0)	37.9 (36.8–38.5)	0.265
SBP, mmHg	82 (71–90)	84 (73–90)	81 (70–90)	0.707
DBP, mmHg	50 (41–56)	50 (42–60)	48 (40–55)	0.313
Heart rate (HR), beats per minute (bpm)	110 (80–129)	111 (82–122)	109 (75–130)	0.952
Respiratory rate (RR), breaths per minute	22 (20–25)	22 (20–24)	24 (20–25)	0.358
Pulse oximetry oxygen saturation (SpO_2), %	94 (89–98)	91 (88–96)	96 (90–98)	0.100
Comorbidities				
Arterial hypertension, n (%)	39 (64)	12 (55)	27 (69)	0.251
Coronary artery disease, n (%)	8 (13)	2 (9)	6 (16)	0.462
PCI/CABG, n (%)	3 (5)	1 (5)	2 (5)	0.919
Atrial fibrillation, n (%)	24 (39)	8 (36)	16 (41)	0.720
Heart failure, n (%)	9 (15)	2 (9)	7 (18)	0.349
Permanent pacemaker, n (%)	3 (5)	0	3 (8)	0.182
Diabetes mellitus, n (%)	23 (38)	7 (32)	16 (41)	0.476
COPD, n (%)	13 (21)	6 (27)	7 (18)	0.393
Prior stroke, n (%)	8 (13)	3 (14)	5 (13)	0.928
Chronic renal dysfunction, n (%)	3 (5)	2 (9)	1 (3)	0.258
Thyroid disease, n (%)	12 (20)	7 (32)	5 (13)	0.073
Dementia, n (%)	17 (28)	4 (18)	13 (33)	0.205
Sleep apnea syndrome, n (%)	1 (2)	1 (5)	0	0.179
Management				
ICU admission, n (%)	7 (12)	3 (14)	4 (10)	0.691

Table 1. *Cont.*

	All Patients (n = 61)	Survivors (n = 22)	Non-Survivors (n = 39)	p-Value
Clinical risk scores				
GCS	12 (9–15)	13.5 (9.75–15)	10 (7–14)	0.012
SOFA	6 (4–8.5)	4.5 (3.75–6.25)	7 (5–9)	0.003
NEWS	12 (10–13.5)	11.5 (10–12.25)	12 (9–14)	0.468
Length of stay (LOS)				
Total LOS, days	6 (2–11)	11 (6–14)	3 (1–8)	<0.001
In-hospital mortality, n (%)	39 (64)	NA	NA	

Abbreviations: CABG = coronary artery bypass grafting; COPD = chronic obstructive pulmonary disease; DBP = diastolic blood pressure; GCS = Glasgow Coma Scale; ICU = intensive care unit; NA = not applicable; NEWS = National Early Warning Score; PCI = percutaneous coronary intervention; SBP = systolic blood pressure; SOFA = Sequential Organ Failure Assessment.

Table 2. Laboratory parameters of study patients upon presentation to the Emergency Department.

	Normal Values	All Patients (n = 61)	Survivors (n = 22)	Non-Survivors (n = 39)	p-Value
Arterial blood gas analysis					
pH	7.350–7.450	7.31 (7.22–7.41)	7.31 (7.22–7.42)	7.31 (7.20–7.41)	0.721
PCO_2, mmHg	32.0–48.0	39 (31–51)	43 (32–57)	38 (30–48)	0.309
HCO_3, mEq/L	22.0–26.0	20 (15–26)	22 (17–28)	19 (15–26)	0.316
Lactate, mmol/L	1.0–1.8	3.6 (2.1–6.8)	3.1 (1.7–4.4)	3.9 (2.6–8)	0.054
Laboratory analysis					
Hemoglobin, g/dL	13.5–17.5	11.3 (9.5–13.3)	11.9 (9.6–14.1)	10.9 (9.3–13)	0.223
WBC, $\times 10^3/\mu L$	4.00–11.00	14.04 (9.55–21.54)	13.62 (8.8–26.22)	14.92 (10.73–21.14)	0.975
PLT, $\times 10^3/\mu L$	150–400	208 (144–329)	204 (151–327)	217 (142–338)	0.866
Urea, mg/dL	16.6–48.5	98 (51–152)	80 (40–113)	109 (63–176)	0.054
Creatinine on admission, mg/dL	0.5–0.9	1.7 (1.0–2.9)	1.5 (0.8–2.5)	2.2 (1.1–3.3)	0.133
Creatinine at 48 h, mg/dL		1.1 (0.8–2.3)	1.0 (0.6–1.8)	1.2 (0.8–2.6)	0.212
Glucose, mg/dL	74–106	155 (102–178)	123 (96–175)	158 (110–179)	0.190
AST, U/L	<32	31 (17–49)	17 (13–34)	33 (25–59)	0.007
ALT, U/L	<33	19 (11–31)	12 (9–26)	23 (12–31)	0.038
CRP, mg/L	0.00–6.00	122 (69–185)	134 (36–196)	119 (79–171)	0.794
Hs-TnT, pg/mL	<14	78 (48–146)	76 (38–132)	80 (52–150)	0.335
Procalcitonin (PCT), ng/mL	<0.5	2.06 (0.84–3.49)	1.41 (0.57–2.43)	2.60 (1.59–4.13)	0.036
SuPAR, ng/mL	<4	13.1 (10.0–21.4)	11.4 (8.2–24)	13.4 (10.6–20.4)	0.481
Proenkephalin (PENK), pmol/L	<80	205 (129–425)	150 (73–220)	250 (159–500)	0.004
Fibrinogen, mg/dL	200–400	404 (340–455)	406 (318–467)	401 (358–452)	0.859

Table 2. *Cont.*

	Normal Values	All Patients (n = 61)	Survivors (n = 22)	Non-Survivors (n = 39)	p-Value
Electrocardiogram					
Sinus rhythm, n (%)		45 (74)	17 (77)	28 (72)	0.640
Atrial fibrillation, n (%)		16 (26)	5 (23)	11 (28)	
Left Ventricular Ejection Fraction					
<40%, n (%)		10 (16)	2 (9)	8 (21)	0.247
≥40%, n (%)		51 (84)	20 (91)	31 (79)	

Abbreviations: ALT = alanine aminotransferase; AST = aspartate aminotransferase; CRP = C-reactive protein; HCO3 = bicarbonate; Hs-TnT = high sensitivity troponin T; PCO2 = partial pressure of carbon dioxide; PLT = platelet count; SuPAR = soluble urokinase plasminogen activator receptor; WBC = white blood cell count.

3.2. Correlations

Analysis revealed significant correlations of PENK with metabolic, renal and sepsis markers. Specifically, LogPENK was positively correlated with LogLactate (rho = 0.369, $p = 0.004$), LogCreatinine (rho = 0.537, $p < 0.001$), LogProcalcitonin (rho = 0.557, $p < 0.001$), and LogSuPAR (rho = 0.327, $p = 0.011$) (Figures 1–4). Additionally, a significant positive correlation was observed between PENK and SOFA score, suggestive of the association of PENK with the patient's clinical severity (rho = 0.391, $p = 0.002$) (Figure 5).

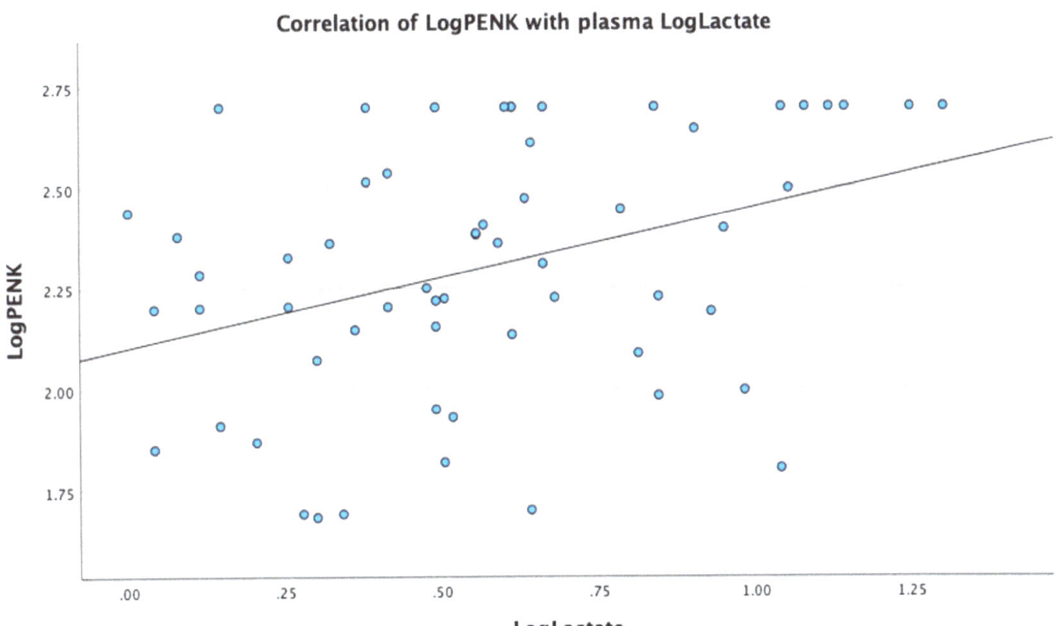

Figure 1. Scatterplot showing the correlation of point-of-care proenkephalin (PENK) with plasma lactate.

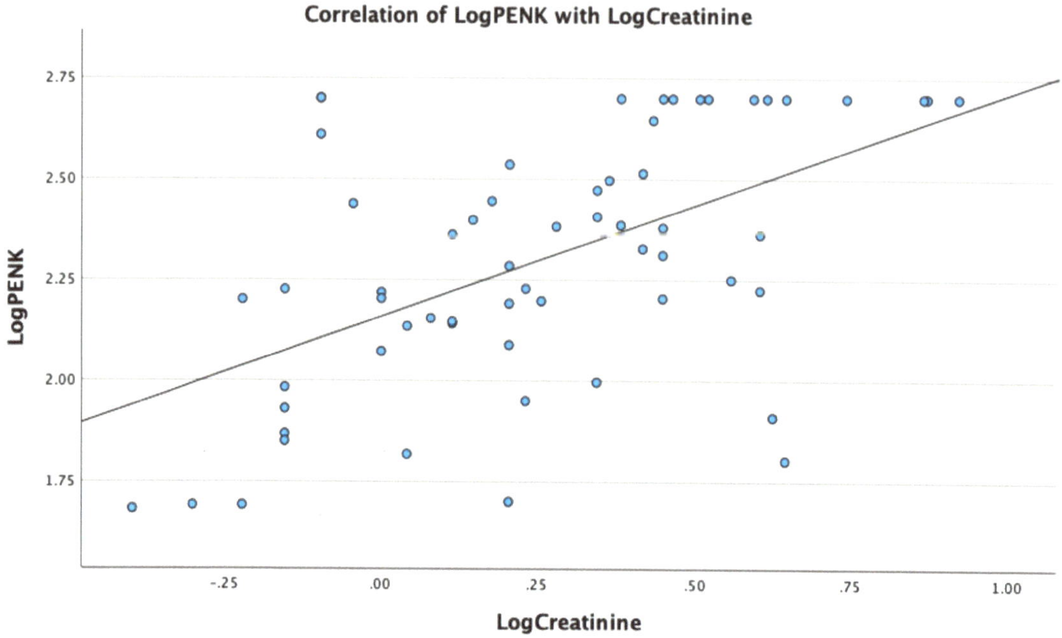

Figure 2. Scatterplot showing the correlation of point-of-care proenkephalin (PENK) with creatinine on admission.

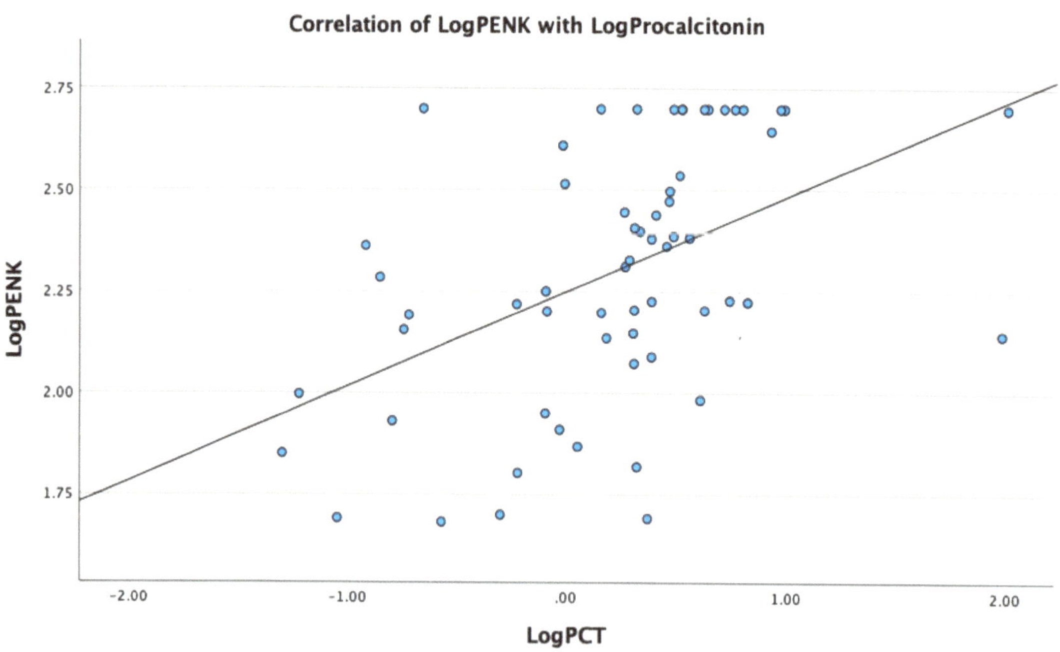

Figure 3. Scatterplot showing the correlation of point-of-care proenkephalin (PENK) with procalcitonin (PCT).

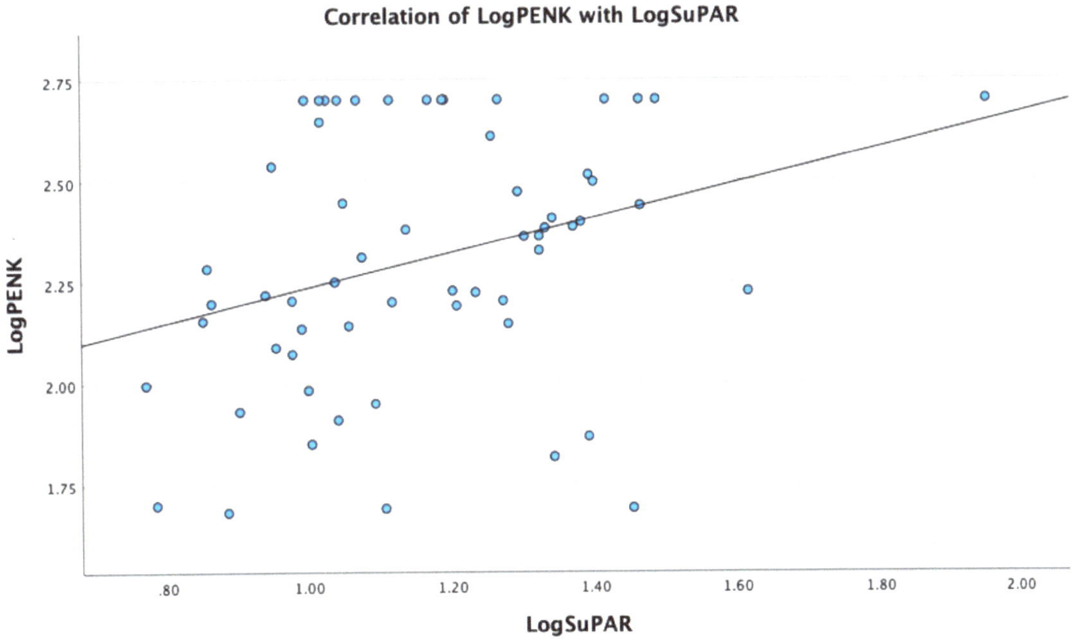

Figure 4. Scatterplot showing the correlation of point-of-care proenkephalin (PENK) with soluble urokinase plasminogen activator receptor (SuPAR).

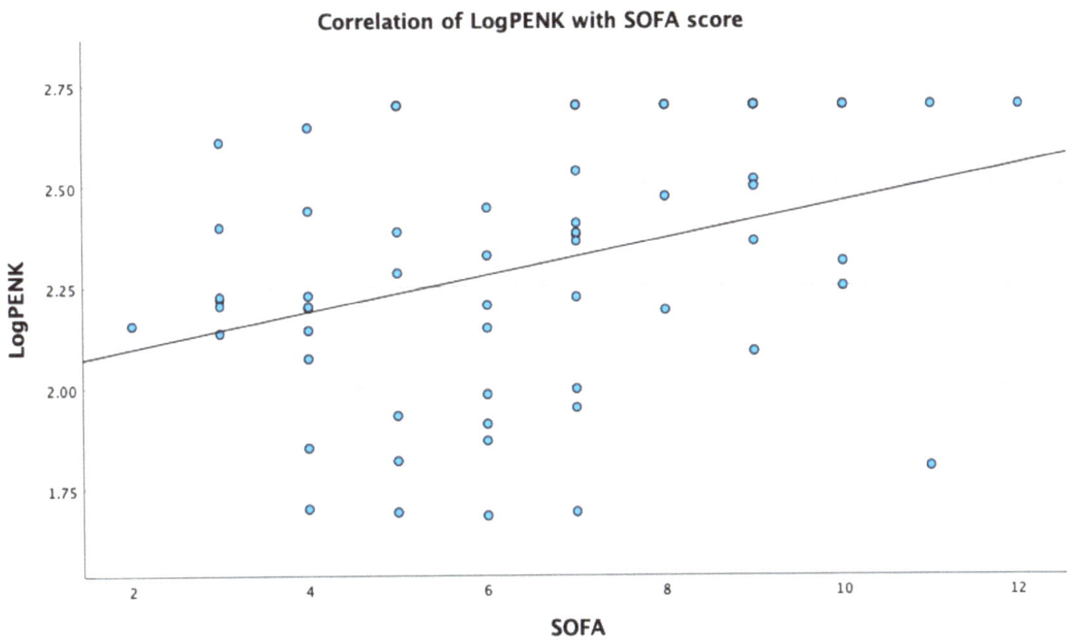

Figure 5. Scatterplot showing the correlation of point-of-care proenkephalin (PENK) with sequential organ failure assessment (SOFA) score.

3.3. Outcomes

During a median (IQR) hospital stay of 6 (2–11) days, 7 patients (12%) developed AKI and 39 (64%) died. POC PENK levels were significantly higher in patients who died [median (IQR), 250 (159–500) pmol/L] compared to survivors [150 (73–220) pmol/L] (p = 0.004) (Figure 6).

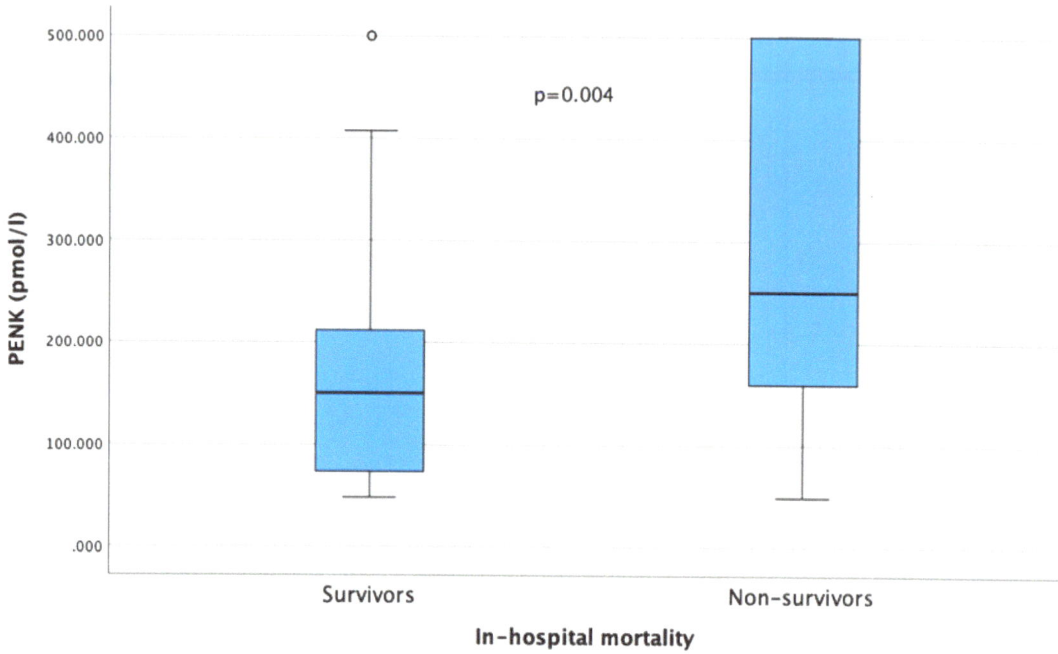

Figure 6. Comparison of point-of-care proenkephalin (PENK) levels between survivors and non-survivors (p = 0.004).

POC PENK levels did not differ significantly between patients with versus without AKI during hospitalization (p > 0.05). In univariate analysis, age [odds ratio (OR) 1.047, 95% confidence interval (CI): 1.002–1.094, p = 0.042], logLactate [OR 6.740, 95% CI: 1.055–43.062, p = 0.044], and logPENK [OR 15.327, 95% CI:2.214–106.094, p = 0.006] were significantly associated with in-hospital mortality. In ROC analysis, POC PENK showed a fair predictive ability for in-hospital mortality, with an area under the curve (AUC) of 0.723 (p = 0.004) (Figure 7). A cut-off of 83.1 pmol/l could discriminate survivors from non-survivors with 95% sensitivity and 32% specificity. In a multivariable logistic regression model that included age, SBP, WBC, logCreatinine, logLactate and logPENK, logPENK was the only independent predictor of in-hospital mortality (OR 11.9, 95% CI: 1.7–84.6, p = 0.013). When diabetes mellitus and arterial hypertension were added in the model, both diabetes (OR 5.3, 95% CI: 1.0–27.6, p = 0.047) and logPENK (OR 12.9, 95% CI: 1.2–136.7, p = 0.034) were independently associated with in-hospital mortality, though logPENK remained the strongest predictor.

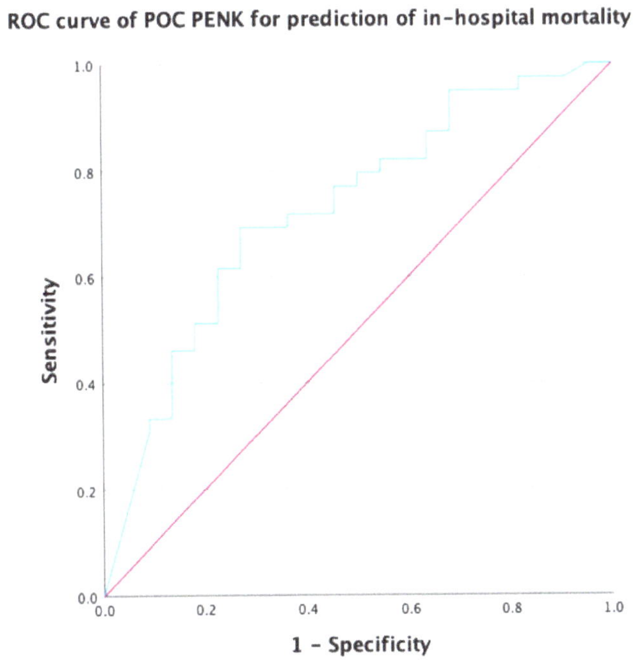

Figure 7. Receiver operating curve (ROC) of point-of-care proenkephalin (PENK) for the prediction of in-hospital mortality. ROC curve is represented by the blue line, whereas the red line corresponds to the diagonal reference line.

4. Discussion

In this prospective observational study, which included unselected patients presenting to the ED with septic shock, we observed that PENK levels upon presentation were significantly associated with metabolic, inflammatory and renal biomarkers and could independently predict in-hospital mortality.

The median value of PENK in our patient cohort with septic shock was 205 (129–425) pmol/L, which is considerably elevated compared to the median PENK concentration reported in healthy subjects (62.3 pmol/L) [16]. The critical clinical condition of our population, as evidenced by the elevated median SOFA score, which is indicative of severe multi-organ dysfunction, may justify the markedly increased concentration of PENK. Likewise, prior studies in septic patients have shown that PENK levels were associated with the severity of sepsis, and were profoundly elevated in the group of patients with septic shock in comparison to patients with sepsis. These studies have reported various median PENK concentrations in patients with septic shock, ranging from 92 (56–171) pmol/L, $p = 0.0001$ [27], to 118.7 (71.7–245.3) pmol/L, $p = 0.02$ [28], and 205 (87–485) pmol/L, $p < 0.001$ [17].

PENK was associated with kidney function, as depicted by a significant correlation of log-transformed baseline PENK values with creatinine levels. This finding is in line with previous studies, in which PENK levels measured in patients with sepsis or septic shock were inversely correlated to creatinine clearance [17] and eGFR [28,29], and positively correlated to creatinine levels [27] and renal SOFA sub-scores [28]. Although serum creatinine is currently used for the assessment of kidney function and the definition and staging of AKI [26], its poor sensitivity and specificity underscore the ever growing need for novel biomarkers that would reflect kidney function in a more precise and timely manner [30].

In septic states, AKI may manifest as a result of perturbations in the renal microvasulature, intrarenal shunting induced by redistribution of blood flow away from the medulla and towards the renal cortex, in combination with focal tubular injury provoked by toxic inflammatory cytokines and oxidative stress leading to apoptosis [31,32]. Recent publications propose the measurement of new early biomarkers of AKI such as Liver Fatty Acid Binding Protein (L-FABP), interleukin-18 (IL-18), Tissue Inhibitor Metal Proteinase-2* Insulin Growth Factor Binding Protein-7 (TIMP2*IGFBP7), Mid-Regional pro-Adrenomedullin (MR-proADM), Kidney Injury Molecule-1 (KIM-1), and Neutrophil Gelatinase-Associated Lipocalin (NGAL) [33–36]. Further investigations are warranted in order to elucidate the usefulness of each biomarker in the context of sepsis-induced AKI.

Increasing evidence suggests that PENK may serve as a functional biomarker of renal function and kidney injury [15,18]; nevertheless, its appropriateness remains to be further elucidated in large-scale clinical trials. In addition, recent studies have shown that PENK levels may predict AKI in patients with sepsis [17,27–29,37,38]. In our population, baseline PENK levels upon ED presentation were not associated with the incidence of AKI. Potential explanations for this observed dissociation between PENK levels and prediction of AKI might be the relatively small patient sample size in our study, which was not powered to examine AKI, coupled with the increased median age in our patient cohort, which might have affected the results, given that most elderly subjects have pre-existing renal decline. Moreover, the high mortality rate within the first 24 h from admission (16/61 patients) eventually resulted in a considerable number of missing values from repeated serum creatinine measurements that would need to be obtained in order to assess AKI in our overall cohort. Additionally, it is worth mentioning a study by Kim et al., which evaluated patients by dividing them into three distinct subgroups, namely suspected sepsis, sepsis and septic shock. In this study, associations between PENK levels and renal function were drawn for each corresponding group of patients. Although PENK levels were positively correlated with the occurrence of AKI, patients with septic shock exhibited elevated PENK levels, regardless of the presence of AKI. In particular, in the subgroup of patients with septic shock, PENK concentrations were 199.7 (101.7–304.3) pmol/L when AKI was present and 117.6 [80.6–209.6] pmol/L in the absence of AKI, yet this difference was not statistically significant ($p = 0.2128$) [29]. Similarly, in the current study, which comprised only patients with septic shock, PENK levels were not significantly associated with the development of AKI.

The aforementioned finding implies that, in the setting of septic shock, other factors, apart from kidney injury, may affect PENK levels. Indeed, in our cohort, baseline PENK levels were also significantly correlated with inflammatory biomarkers, namely procalcitonin and SuPAR, as well as with lactate, which is regarded as a marker of metabolic dysregulation. SuPAR is a glycoprotein mainly expressed on various immune cells and released in the bloodstream upon an inflammatory insult [39]. Elevated levels are considered to reflect the degree of inflammation and are associated with disease severity, while they may also aid in the risk stratification of septic patients presenting to the ED [40–42]. In parallel, procalcitonin has been established as a new biomarker for the diagnosis and prognosis of patients with sepsis caused by bacterial infections [43,44]. Regarding the role of lactate in sepsis, hyperlactemia is deemed to be the result of tissue hypoxia combined with overt stress response [45]. Our findings seem to corroborate the notion that elevation of PENK levels in patients with septic shock is multifactorial. In fact, it could be assumed that overactivation of the inflammatory system coupled with hemodynamic compromise and adrenergic overdrive may play a significant role either in the upregulation or even in the decreased clearance of PENK. A possible underlying mechanism observed in an experimental study which, however, warrants extensive testing and further elucidation, is the upregulation of PENK induced by oxidative stress which is known to prevail in septic states and further complicate the multifaceted pathobiology of sepsis [46,47]. The complex interplay between septic shock and increased PENK levels is depicted in Figure 8, where plausible pathophysiological mechanisms that may come into play are being proposed, highlighting the multifactorial interaction between septic shock and PENK.

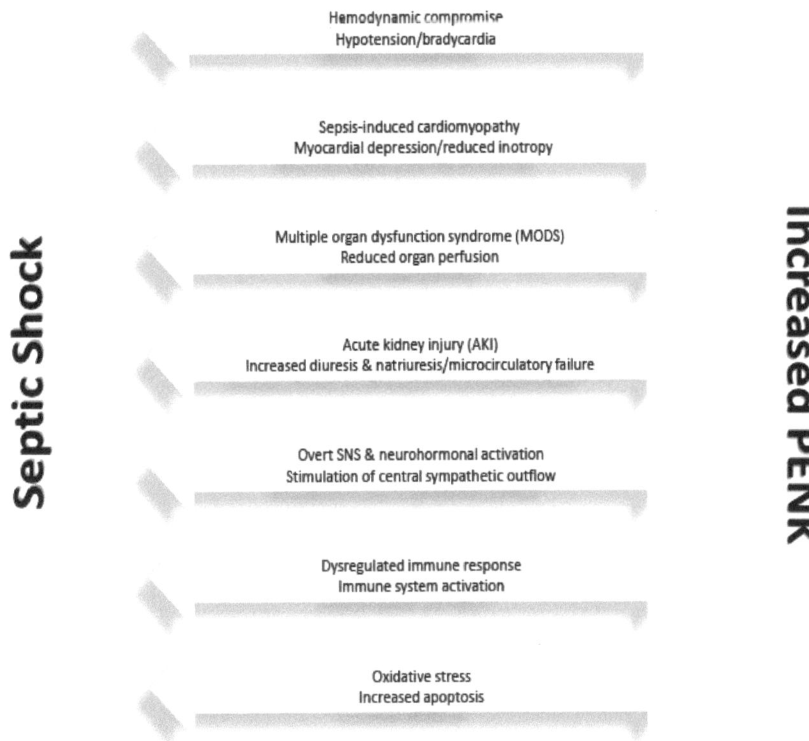

Figure 8. The complex interplay between septic shock and increased proenkephalin levels. Septic shock is characterized by profound hemodynamic instability, which can be further aggravated by the development of sepsis-induced cardiomyopathy [48]. On the other hand, elevated proenkephalin (PENK) levels can lead to further hemodynamic compromise, since it has been shown that PENK induces hypotension and bradycardia, both of which may exert deleterious effects on an already dysregulated cardiovascular system. Decreased blood pressure and heart rate have been attributed to PENK's inhibitory effect on the catecholaminergic neural stimulation of the heart and the vasculature, resulting in an attenuated response of the heart to adrenergic stimulation, along with peripheral vasodilation through diminished response to sympathetic vasoconstriction. Moreover, increased PENK levels can exacerbate cardiac dysfunction by provoking myocardial depression and causing a direct negative inotropic effect [49–51]. Additionally, septic shock can lead to multiple organ dysfunction syndrome (MODS), which can be further mediated by elevated PENK levels via decreased organ perfusion [49,52]. Acute kidney injury (AKI) represents a common finding in patients with septic shock, whereas PENK has been shown to induce natriuresis and diuresis through delta (δ) opioid receptors [25,53]. In this regard, it has been postulated that elevated PENK levels in patients with AKI act as a counter-regulatory mechanism in an attempt to mitigate worsening renal function which, however, becomes maladaptive, eventually leading to untoward alterations in renal excretory function, regional blood flow redistribution, microcirculatory failure and further decline in renal perfusion [17,27,37,54]. Apart from that, a hallmark of septic shock is the overt activation of the sympathetic nervous system (SNS) and the neurohormonal system. In parallel, increased PENK levels have been implicated in the stimulation of the central sympathetic outflow, thus leading to further sympathetic overdrive which, however, may subsequently result in sympathetic exhaustion and downregulation of the sympathetic pathway at the cardiac level [55]. Dysregulated immune response is another key feature of septic shock, while PENK is known to be involved in immunomodulation.

Accordingly, elevated PENK levels result in increased release of cytokines by macrophages and T-cells, coupled with marked proliferation of B and T lymphocytes and augmented activation of macrophages, natural killer (NK) cells and neutrophils. The net result is overactivation and exaggerated dysregulation of the immune system [14,56]. Finally, septic shock is governed by tissue hypoxia, leading to cellular dysfunction, mitochondrial dysfunction, metabolic dysregulation and abnormal calcium handling. These abnormalities collectively result in increased oxidative stress and increased production of reactive oxygen species (ROS), which are further aggravated by the proapoptotic effects of increased PENK levels, ultimately leading to a severe compromise of cellular structure and function [46,57]. It should be noted that all of the above represent postulated mechanisms that are potentially implicated in the multifaceted interaction between septic shock and elevated PENK levels.

Importantly, in the current study, PENK levels upon presentation were the strongest predictor of in-hospital mortality (OR 11.9, 95% CI: 1.7–84.6, $p = 0.013$). Similar findings have been reported by previous trials [17,27–29,37,38]. PENK levels differed significantly ($p = 0.004$) between survivors [median (IQR), 150 (73–220) pmol/L] and non-survivors [250 (159–500) pmol/L], thus implying that PENK may serve as an important biomarker for the early risk stratification of patients presenting to the ED with septic shock. This finding becomes even more important, considering the fact that the patients in our study had features of severe form of septic shock, as suggested by high SOFA score [58] and high in-hospital mortality (64%, with 26% dying within the first 24 h after admission). Therefore, it could be postulated that even among patients lying in the severe extremes of the septic shock spectrum, PENK measured upon ED presentation may discriminate those with the highest mortality risk. By doing so, POC PENK measurement upon ED arrival may substantially aid in the decision-making process with regard to the provision of appropriate type of treatment, level of monitoring and patient allocation. Although patient mortality in our study was strikingly high (64%), when compared to previously published studies examining the prognostic utility of PENK in patients with sepsis, it has to be emphasized that we recruited unselected elderly patients presenting to the ED exclusively with septic shock, and we measured in-hospital mortality, contrary to prior studies that measured 7-day [17], 30-day [28,29,37,38] and 90-day [27] mortality in patients recruited in the ED [17,28,38] and the ICU [27–29,37]. Certain baseline patient characteristics, such as the high median age (83 years) and the presence of multiple comorbidities, coupled with a high SOFA score [58], may in part explain both the high in-hospital mortality in the overall population and the notable percentage of patients (26%) who died within the first 24 h from ED admission (comprising 41% of the total deaths).

An important strength of this study is that PENK levels were measured upon presentation to the ED at the patient's bedside by means of a point-of-care device, using whole blood sample. This measurement method of POC PENK levels yielded statistically significant results regarding survival prediction. Our study points towards the fact that previously reported results regarding the prognostic role of PENK levels, from studies utilizing frozen plasma [17,27–29,37,38] of patients with septic shock, are reproducible with a point of care device in the ED. This is of major importance, considering the fact that the initial resuscitation of patients with septic shock takes place in the ED, where readily available PENK levels on-site may provide additional information during ongoing resuscitation. The recent literature suggests that patients with sepsis may face significant admission delays to the ICU, a phenomenon that inflicts an additional mortality burden in the already compromised subgroup of patients with septic shock [59]. Risk stratification, provided by a single PENK value upon ED presentation, may identify a subset of patients who carry the poorest prognosis, and may thus benefit from the implementation of an aggressive therapy protocol in combination with close monitoring and early ICU admission.

Limitations

Undoubtedly, the small population of our study is a significant limitation, yet the current findings may serve as a boost for further research on the role of PENK in septic shock states, as well as on the pathophysiological pathways responsible for its increase in patients with septic shock. More importantly, our survey suggests that PENK levels may distinguish survivors from non-survivors; this may actually act as an impetus for future studies to further focus on determining a cut-off level that would discriminate patients with dismal survival from those with favorable prognosis. Moreover, we examined the potential utility of a single measurement of PENK upon patient presentation to the ED that could aid in the risk stratification of patients with septic shock. Future studies could further evaluate the prognostic importance of changes in PENK levels after admission.

5. Conclusions

In conclusion, POC PENK measured upon presentation to the ED may reliably and independently predict poor survival among patients with septic shock in a timely manner. The findings of this study highlight an emerging prognostic role of PENK as a promising POC biomarker in the ED setting, which may facilitate in the early identification of subjects with septic shock who require close monitoring and intensive care management.

Supplementary Materials: The following supporting information can be downloaded at: https://www.mdpi.com/article/10.3390/biomedicines12051004/s1. Table S1: Sepsis and septic shock clinical criteria according to the Third International Consensus Definitions for Sepsis and Septic Shock, modified from [1]; Table S2: Proposed staging for acute kidney injury according to KDIGO criteria, modified from [26].

Author Contributions: Conceptualization, E.P. and J.P.; methodology, C.V. and V.B.; formal analysis, V.B.; data curation, D.M. and C.V.; writing—original draft preparation, C.V.; writing—review and editing, S.B., D.M., V.B. and I.V.; supervision, S.T. All authors have read and agreed to the published version of the manuscript.

Funding: This research received no external funding.

Institutional Review Board Statement: The study was conducted in accordance with the Declaration of Helsinki, and approved by the Institutional Review Board (or Ethics Committee) of ATTIKON UNIVERSITY HOSPITAL (EBΔ282/10-05-2022).

Informed Consent Statement: Informed consent was obtained from all subjects involved in the study.

Data Availability Statement: Data are contained within article.

Conflicts of Interest: C.V., S.B., V.B., I.V., D.M., S.T. and E.P declare no conflicts of interest. J.P has received a speaker honorarium from Roche and Sphingotec.

References

1. Singer, M.; Deutschman, C.S.; Seymour, C.W.; Shankar-Hari, M.; Annane, D.; Bauer, M.; Bellomo, R.; Bernard, G.R.; Chiche, J.-D.; Coopersmith, C.M.; et al. The Third International Consensus Definitions for Sepsis and Septic Shock (Sepsis-3). *JAMA* **2016**, *315*, 801. [CrossRef] [PubMed]
2. Rudd, K.E.; Johnson, S.C.; Agesa, K.M.; Shackelford, K.A.; Tsoi, D.; Kievlan, D.R.; Colombara, D.V.; Ikuta, K.S.; Kissoon, N.; Finfer, S.; et al. Global, Regional, and National Sepsis Incidence and Mortality, 1990–2017: Analysis for the Global Burden of Disease Study. *Lancet* **2020**, *395*, 200–211. [CrossRef] [PubMed]
3. Fleischmann, C.; Scherag, A.; Adhikari, N.K.J.; Hartog, C.S.; Tsaganos, T.; Schlattmann, P.; Angus, D.C.; Reinhart, K. Assessment of Global Incidence and Mortality of Hospital-Treated Sepsis. Current Estimates and Limitations. *Am. J. Respir. Crit. Care Med.* **2016**, *193*, 259–272. [CrossRef] [PubMed]
4. Sakr, Y.; Jaschinski, U.; Wittebole, X.; Szakmany, T.; Lipman, J.; Ñamendys-Silva, S.A.; Martin-Loeches, I.; Leone, M.; Lupu, M.-N.; Vincent, J.-L.; et al. Sepsis in Intensive Care Unit Patients: Worldwide Data From the Intensive Care over Nations Audit. *Open Forum Infect. Dis.* **2018**, *5*, ofy313. [CrossRef] [PubMed]
5. Dyck, B.; Unterberg, M.; Adamzik, M.; Koos, B. The Impact of Pathogens on Sepsis Prevalence and Outcome. *Pathogens* **2024**, *13*, 89. [CrossRef] [PubMed]
6. Martin, G.S. Sepsis, Severe Sepsis and Septic Shock: Changes in Incidence, Pathogens and Outcomes. *Expert Rev. Anti-Infect. Ther.* **2012**, *10*, 701–706. [CrossRef] [PubMed]
7. Arora, J.; Mendelson, A.A.; Fox-Robichaud, A. Sepsis: Network Pathophysiology and Implications for Early Diagnosis. *Am. J. Physiol.-Regul. Integr. Comp. Physiol.* **2023**, *324*, R613–R624. [CrossRef] [PubMed]

8. Jarczak, D.; Kluge, S.; Nierhaus, A. Sepsis—Pathophysiology and Therapeutic Concepts. *Front. Med.* **2021**, *8*, 628302. [CrossRef] [PubMed]
9. Vincent, J.-L.; Moreno, R.; Takala, J.; Willatts, S.; De Mendonça, A.; Bruining, H.; Reinhart, C.K.; Suter, P.M.; Thijs, L.G. The SOFA (Sepsis-Related Organ Failure Assessment) Score to Describe Organ Dysfunction/Failure: On Behalf of the Working Group on Sepsis-Related Problems of the European Society of Intensive Care Medicine (See Contributors to the Project in the Appendix). *Intensive Care Med.* **1996**, *22*, 707–710. [CrossRef]
10. Yao, L.; Zhang, L.; Zhou, C. Analysis of Prognostic Risk Factors of Sepsis Patients in Intensive Care Unit Based on Data Analysis. *J. Healthc. Eng.* **2022**, *2022*, 3746640. [CrossRef]
11. Girard, T.D.; Opal, S.M.; Ely, E.W. Insights into Severe Sepsis in Older Patients: From Epidemiology to Evidence-Based Management. *Clin. Infect. Dis.* **2005**, *40*, 719–727. [CrossRef] [PubMed]
12. Alobaidi, R.; Basu, R.K.; Goldstein, S.L.; Bagshaw, S.M. Sepsis-Associated Acute Kidney Injury. *Semin. Nephrol.* **2015**, *35*, 2–11. [CrossRef]
13. Siew, E.D.; Fissell, W.H.; Tripp, C.M.; Blume, J.D.; Wilson, M.D.; Clark, A.J.; Vincz, A.J.; Ely, E.W.; Pandharipande, P.P.; Girard, T.D. Acute Kidney Injury as a Risk Factor for Delirium and Coma during Critical Illness. *Am. J. Respir. Crit. Care Med.* **2017**, *195*, 1597–1607. [CrossRef] [PubMed]
14. Denning, G.M.; Ackermann, L.W.; Barna, T.J.; Armstrong, J.G.; Stoll, L.L.; Weintraub, N.L.; Dickson, E.W. Proenkephalin Expression and Enkephalin Release Are Widely Observed in Non-Neuronal Tissues. *Peptides* **2008**, *29*, 83–92. [CrossRef]
15. Beunders, R.; Struck, J.; Wu, A.H.B.; Zarbock, A.; Di Somma, S.; Mehta, R.L.; Koyner, J.L.; Nadim, M.K.; Maisel, A.S.; Murray, P.T.; et al. Proenkephalin (PENK) as a Novel Biomarker for Kidney Function. *J. Appl. Lab. Med.* **2017**, *2*, 400–412. [CrossRef] [PubMed]
16. Ernst, A.; Köhrle, J.; Bergmann, A. Proenkephalin A 119–159, a Stable Proenkephalin A Precursor Fragment Identified in Human Circulation. *Peptides* **2006**, *27*, 1835–1840. [CrossRef] [PubMed]
17. Marino, R.; Struck, J.; Hartmann, O.; Maisel, A.S.; Rehfeldt, M.; Magrini, L.; Melander, O.; Bergmann, A.; Di Somma, S. Diagnostic and Short-Term Prognostic Utility of Plasma pro-Enkephalin (pro-ENK) for Acute Kidney Injury in Patients Admitted with Sepsis in the Emergency Department. *J. Nephrol.* **2015**, *28*, 717–724. [CrossRef] [PubMed]
18. Khorashadi, M.; Beunders, R.; Pickkers, P.; Legrand, M. Proenkephalin: A New Biomarker for Glomerular Filtration Rate and Acute Kidney Injury. *Nephron* **2020**, *144*, 655–661. [CrossRef] [PubMed]
19. Emmens, J.E.; Ter Maaten, J.M.; Damman, K.; van Veldhuisen, D.J.; de Boer, R.A.; Struck, J.; Bergmann, A.; Sama, I.E.; Streng, K.W.; Anker, S.D.; et al. Proenkephalin, an Opioid System Surrogate, as a Novel Comprehensive Renal Marker in Heart Failure. *Circ. Heart Fail.* **2019**, *12*, e005544. [CrossRef]
20. Shankar-Hari, M.; Phillips, G.S.; Levy, M.L.; Seymour, C.W.; Liu, V.X.; Deutschman, C.S.; Angus, D.C.; Rubenfeld, G.D.; Singer, M.; for the Sepsis Definitions Task Force. Developing a New Definition and Assessing New Clinical Criteria for Septic Shock: For the Third International Consensus Definitions for Sepsis and Septic Shock (Sepsis-3). *JAMA* **2016**, *315*, 775. [CrossRef]
21. Teasdale, G.; Jennett, B. ASSESSMENT OF COMA AND IMPAIRED CONSCIOUSNESS. *Lancet* **1974**, *304*, 81–84. [CrossRef] [PubMed]
22. Levey, A.S.; Coresh, J.; Greene, T.; Marsh, J.; Stevens, L.A.; Kusek, J.W.; Van Lente, F.; Chronic Kidney Disease Epidemiology Collaboration. Expressing the Modification of Diet in Renal Disease Study Equation for Estimating Glomerular Filtration Rate with Standardized Serum Creatinine Values. *Expressing the Modification of Diet in Renal Disease Study Equation for Estimating Glomerular Filtration Rate with Standardized Serum Creatinine Values. Clin. Chem.* **2007**, *53*, 766–772. [CrossRef] [PubMed]
23. Hawkes, N. Royal College Recommends National System to Recognise Deteriorating Patients. *BMJ* **2012**, *345*, e5041. [CrossRef] [PubMed]
24. Nexus IB10 | NexusDx. Available online: https://www.nexus-dx.com/nexus-ib10/ (accessed on 20 April 2024).
25. Donato, L.J.; Meeusen, J.W.; Lieske, J.C.; Bergmann, D.; Sparwaßer, A.; Jaffe, A.S. Analytical Performance of an Immunoassay to Measure Proenkephalin. *Clin. Biochem.* **2018**, *58*, 72–77. [CrossRef] [PubMed]
26. Khwaja, A. KDIGO Clinical Practice Guidelines for Acute Kidney Injury. *Nephron Clin. Pract.* **2012**, *120*, c179–c184. [CrossRef] [PubMed]
27. Caironi, P.; Latini, R.; Struck, J.; Hartmann, O.; Bergmann, A.; Bellato, V.; Ferraris, S.; Tognoni, G.; Pesenti, A.; Gattinoni, L.; et al. Circulating Proenkephalin, Acute Kidney Injury, and Its Improvement in Patients with Severe Sepsis or Shock. *Clin. Chem.* **2018**, *64*, 1361–1369. [CrossRef] [PubMed]
28. Kim, H.; Hur, M.; Struck, J.; Bergmann, A.; Somma, S.D.; GREAT Network. Proenkephalin Predicts Organ Failure, Renal Replacement Therapy, and Mortality in Patients With Sepsis. *Ann. Lab. Med.* **2020**, *40*, 466–473. [CrossRef]
29. Kim, H.; Hur, M.; Lee, S.; Marino, R.; Magrini, L.; Cardelli, P.; Struck, J.; Bergmann, A.; Hartmann, O.; Somma, S.D.; et al. Proenkephalin, Neutrophil Gelatinase-Associated Lipocalin, and Estimated Glomerular Filtration Rates in Patients With Sepsis. *Ann. Lab. Med.* **2017**, *37*, 388–397. [CrossRef] [PubMed]
30. Endre, Z.H.; Pickering, J.W.; Walker, R.J. Clearance and beyond: The Complementary Roles of GFR Measurement and Injury Biomarkers in Acute Kidney Injury (AKI). *Am. J. Physiol.-Ren. Physiol.* **2011**, *301*, F697–F707. [CrossRef]
31. Bellomo, R.; Kellum, J.A.; Ronco, C.; Wald, R.; Martensson, J.; Maiden, M.; Bagshaw, S.M.; Glassford, N.J.; Lankadeva, Y.; Vaara, S.T.; et al. Acute Kidney Injury in Sepsis. *Intensive Care Med.* **2017**, *43*, 816–828. [CrossRef]
32. Post, E.H.; Kellum, J.A.; Bellomo, R.; Vincent, J.-L. Renal Perfusion in Sepsis: From Macro- to Microcirculation. *Kidney Int.* **2017**, *91*, 45–60. [CrossRef] [PubMed]

33. Moledina, D.G.; Parikh, C.R. Phenotyping of Acute Kidney Injury: Beyond Serum Creatinine. *Semin. Nephrol.* **2018**, *38*, 3–11. [CrossRef] [PubMed]
34. Forni, L.G.; Joannidis, M.; Artigas, A.; Bell, M.; Hoste, E.; Joannes-Boyau, O.; Kashani, K.; Koyner, J.; Kimmele, T.; Shi, J.; et al. Characterising Acute Kidney Injury: The Complementary Roles of Biomarkers of Renal Stress and Renal Function. *J. Crit. Care* **2022**, *71*, 154066. [CrossRef] [PubMed]
35. Zou, C.; Wang, C.; Lu, L. Advances in the Study of Subclinical AKI Biomarkers. *Front. Physiol.* **2022**, *13*, 960059. [CrossRef] [PubMed]
36. Lacquaniti, A.; Ceresa, F.; Campo, S.; Barbera, G.; Caruso, D.; Palazzo, E.; Patanè, F.; Monardo, P. Acute Kidney Injury and Sepsis after Cardiac Surgery: The Roles of Tissue Inhibitor Metalloproteinase-2, Insulin-like Growth Factor Binding Protein-7, and Mid-Regional Pro-Adrenomedullin. *JCM* **2023**, *12*, 5193. [CrossRef] [PubMed]
37. Hollinger, A.; Wittebole, X.; François, B.; Pickkers, P.; Antonelli, M.; Gayat, E.; Chousterman, B.G.; Lascarrou, J.-B.; Dugernier, T.; Di Somma, S.; et al. Proenkephalin A 119-159 (Penkid) Is an Early Biomarker of Septic Acute Kidney Injury: The Kidney in Sepsis and Septic Shock (Kid-SSS) Study. *Kidney Int. Rep.* **2018**, *3*, 1424–1433. [CrossRef] [PubMed]
38. Rosenqvist, M.; Bronton, K.; Hartmann, O.; Bergmann, A.; Struck, J.; Melander, O. Proenkephalin a 119–159 (penKid)—A Novel Biomarker for Acute Kidney Injury in Sepsis: An Observational Study. *BMC Emerg. Med.* **2019**, *19*, 75. [CrossRef] [PubMed]
39. Desmedt, S.; Desmedt, V.; Delanghe, J.R.; Speeckaert, R.; Speeckaert, M.M. The Intriguing Role of Soluble Urokinase Receptor in Inflammatory Diseases. *Crit. Rev. Clin. Lab. Sci.* **2017**, *54*, 117–133. [CrossRef] [PubMed]
40. Velissaris, D.; Dimopoulos, G.; Parissis, J.; Alexiou, Z.; Antonakos, N.; Babalis, D.; Gerakari, S.; Kaldis, V.; Koutoukas, P.; Lada, M.; et al. Prognostic Role of Soluble Urokinase Plasminogen Activator Receptor at the Emergency Department: A Position Paper by the Hellenic Sepsis Study Group. *Infect. Dis. Ther.* **2020**, *9*, 407–416. [CrossRef]
41. Velissaris, D.; Pierrakos, C.; Karamouzos, V.; Pantzaris, N.D.; Gogos, C. The Use of Soluble Urokinase Plasminogen Activator Receptor (suPAR) as a Marker of Sepsis in the Emergency Department Setting. A Current Review. *Acta Clin. Belg.* **2021**, *76*, 79–84. [CrossRef]
42. Backes, Y.; Van Der Sluijs, K.F.; Mackie, D.P.; Tacke, F.; Koch, A.; Tenhunen, J.J.; Schultz, M.J. Usefulness of suPAR as a Biological Marker in Patients with Systemic Inflammation or Infection: A Systematic Review. *Intensive Care Med.* **2012**, *38*, 1418–1428. [CrossRef]
43. Tan, M.; Lu, Y.; Jiang, H.; Zhang, L. The Diagnostic Accuracy of Procalcitonin and C-reactive Protein for Sepsis: A Systematic Review and Meta-analysis. *J. Cell. Biochem.* **2019**, *120*, 5852–5859. [CrossRef] [PubMed]
44. Kakar, A.; Ray, S.; Mamidipalli, R.; Jain, R.; Ghalaut, M.S.; Choudhury, S. A Comparative Study of the Diagnostic and Prognostic Utility of Soluble Urokinase-Type Plasminogen Activator Receptor and Procalcitonin in Patients with Sepsis and Systemic Inflammation Response Syndrome. *Indian J. Crit. Care Med.* **2020**, *24*, 245–251. [CrossRef]
45. Garcia-Alvarez, M.; Marik, P.; Bellomo, R. Sepsis-Associated Hyperlactatemia. *Crit. Care* **2014**, *18*, 503. [CrossRef]
46. Rosenberger, J.; Petrovics, G.; Buzas, B. Oxidative Stress Induces Proorphanin FQ and Proenkephalin Gene Expression in Astrocytes through P38- and ERK-MAP Kinases and NF-κB. *J. Neurochem.* **2001**, *79*, 35–44. [CrossRef] [PubMed]
47. Gomez, H.; Ince, C.; De Backer, D.; Pickkers, P.; Payen, D.; Hotchkiss, J.; Kellum, J.A. A Unified Theory of Sepsis-Induced Acute Kidney Injury: Inflammation, Microcirculatory Dysfunction, Bioenergetics, and the Tubular Cell Adaptation to Injury. *Shock* **2014**, *41*, 3–11. [CrossRef] [PubMed]
48. Hollenberg, S.M.; Singer, M. Pathophysiology of Sepsis-Induced Cardiomyopathy. *Nat. Rev. Cardiol.* **2021**, *18*, 424–434. [CrossRef]
49. Ng, L.L.; Sandhu, J.K.; Narayan, H.; Quinn, P.A.; Squire, I.B.; Davies, J.E.; Bergmann, A.; Maisel, A.; Jones, D.J.L. Proenkephalin and Prognosis after Acute Myocardial Infarction. *J. Am. Coll. Cardiol.* **2014**, *63*, 280–289. [CrossRef] [PubMed]
50. van den Brink, O.W.V.; Delbridge, L.M.; Rosenfeldt, F.L.; Penny, D.; Esmore, D.S.; Quick, D.; Kaye, D.M.; Pepe, S. Endogenous Cardiac Opioids: Enkephalins in Adaptation and Protection of the Heart. *Heart Lung Circ.* **2003**, *12*, 178–187. [CrossRef]
51. Pepe, S.; van den Brink, O.W.V.; Lakatta, E.G.; Xiao, R.-P. Cross-Talk of Opioid Peptide Receptor and Beta-Adrenergic Receptor Signalling in the Heart. *Cardiovasc. Res.* **2004**, *63*, 414–422. [CrossRef]
52. Frigyesi, A.; Boström, L.; Lengquist, M.; Johnsson, P.; Lundberg, O.H.M.; Spångfors, M.; Annborn, M.; Cronberg, T.; Nielsen, N.; Levin, H.; et al. Plasma Proenkephalin A 119-159 on Intensive Care Unit Admission Is a Predictor of Organ Failure and 30-Day Mortality. *Intensive Care Med. Exp.* **2021**, *9*, 36. [CrossRef] [PubMed]
53. Sezen, S.F.; Kenigs, V.A.; Kapusta, D.R. Renal Excretory Responses Produced by the Delta Opioid Agonist, BW373U86, in Conscious Rats. *J. Pharmacol. Exp. Ther.* **1998**, *287*, 238–245. [PubMed]
54. Ng, L.L.; Squire, I.B.; Jones, D.J.L.; Cao, T.H.; Chan, D.C.S.; Sandhu, J.K.; Quinn, P.A.; Davies, J.E.; Struck, J.; Hartmann, O.; et al. Proenkephalin, Renal Dysfunction, and Prognosis in Patients With Acute Heart Failure: A GREAT Network Study. *J. Am. Coll. Cardiol.* **2017**, *69*, 56–69. [CrossRef] [PubMed]
55. Holaday, J.W. Cardiovascular Effects of Endogenous Opiate Systems. *Annu. Rev. Pharmacol. Toxicol.* **1983**, *23*, 541–594. [CrossRef] [PubMed]
56. Salzet, M.; Tasiemski, A. Involvement of Pro-Enkephalin-Derived Peptides in Immunity. *Dev. Comp. Immunol.* **2001**, *25*, 177–185. [CrossRef]
57. McTavish, N.; Copeland, L.A.; Saville, M.K.; Perkins, N.D.; Spruce, B.A. Proenkephalin Assists Stress-Activated Apoptosis through Transcriptional Repression of NF-kappaB- and P53-Regulated Gene Targets. *Cell Death Differ.* **2007**, *14*, 1700–1710. [CrossRef] [PubMed]

58. Vincent, J.-L.; De Mendonca, A.; Cantraine, F.; Moreno, R.; Takala, J.; Suter, P.M.; Sprung, C.L.; Colardyn, F.; Blecher, S. Use of the SOFA Score to Assess the Incidence of Organ Dysfunction/Failure in Intensive Care Units: Results of a Multicenter, Prospective Study. *Crit. Care Med.* **1998**, *26*, 1793–1800. [CrossRef]
59. Ahn, Y.H.; Lee, J.; Oh, D.K.; Lee, S.Y.; Park, M.H.; Lee, H.; Lim, C.-M.; Lee, S.-M.; Lee, H.Y.; the Korean Sepsis Alliance (KSA) Investigators; et al. Association between the Timing of ICU Admission and Mortality in Patients with Hospital-Onset Sepsis: A Nationwide Prospective Cohort Study. *J. Intensive Care* **2023**, *11*, 16. [CrossRef]

biomedicines

MDPI

Article

Ischemia-Modified Albumin, Lactate, and Combination for Predicting Mortality in Patients with Septic Shock in the Emergency Department

Bo-Yeong Jin [1], Sukyo Lee [2], Woosik Kim [2], Jong-Hak Park [2], Hanjin Cho [2], Sungwoo Moon [2] and Sejoong Ahn [2,*]

1 Department of Biomedical Sciences, Seoul National University College of Medicine, Seoul 03080, Republic of Korea; boyeongjin0@gmail.com
2 Department of Emergency Medicine, Korea University Ansan Hospital, Ansan-si 15355, Republic of Korea
* Correspondence: sejoongahn@naver.com or sejoongahn@gmail.com; Tel.: +82-31-412-5381

Abstract: Ischemia-modified albumin (IMA) is produced during ischemia and reactive oxygen species production. This study aimed to evaluate the association between IMA and mortality in a larger population and the prognostic value of the combination of IMA and lactate for predicting mortality in septic shock patients in the emergency department. This retrospective observational study included adult septic shock patients between October 2019 and December 2021. A multivariable Cox proportional hazards model was performed. IMA was significantly higher in the non-surviving group than in the surviving group (89.1 ± 7.2 vs. 83.8 ± 6.2 U/mL, $p < 0.001$). IMA was independently associated with 28-day mortality after adjustments (adjusted hazard ratio [aHR]: 1.075, 95% confidence interval [CI]: 1.016–1.138, $p = 0.012$). The area under the ROC curve (AUROC) of IMA was 0.712 (95% CI: 0.648–0.775, $p < 0.001$) and was comparable to that of lactate. The AUROC of the combination of IMA and lactate was 0.838 (95% CI: 0.786–0.889, $p < 0.001$). The group with both high lactate and high IMA levels showed an extremely high risk of mortality than other groups (86.1%; aHR 8.956, 95% CI 4.071–19.70, $p < 0.001$). The elevation of IMA was associated with mortality in septic shock patients. The combination of IMA and lactate can be a helpful tool for early risk stratification of septic shock patients.

Keywords: septic shock; ischemia modified albumin; lactic acid; prognosis; emergency department; risk stratification

Citation: Jin, B.-Y.; Lee, S.; Kim, W.; Park, J.-H.; Cho, H.; Moon, S.; Ahn, S. Ischemia-Modified Albumin, Lactate, and Combination for Predicting Mortality in Patients with Septic Shock in the Emergency Department. *Biomedicines* **2024**, *12*, 1421. https://doi.org/10.3390/biomedicines12071421

Academic Editors: Wen-Lin Su, Sheng-Kang Chiu, Chih-Hao Shen and Yi-Ting Chen

Received: 31 May 2024
Revised: 22 June 2024
Accepted: 24 June 2024
Published: 26 June 2024

1. Introduction

Sepsis is a life-threatening organ dysfunction caused by a dysregulated host response to infection [1]. Sepsis is a major global health problem with a high mortality [2,3]. Septic shock is the most severe form of sepsis, with a mortality rate of approximately 35–38% [3,4]. Early recognition, risk stratification, and comprehensive management are important to reduce mortality in patients with sepsis and septic shock, especially in emergency departments.

Several biomarkers have been proposed for predicting the outcomes of sepsis and septic shock. Lactate is a well-known biomarker of cellular metabolic dysfunction [3,5,6]. The elevation of lactate levels can be caused by inadequate tissue oxygen delivery, increased anaerobic metabolism, or increased lactate production [5–7]. Lactate is widely used to assess disease severity and predict mortality in critically ill patients, including those with sepsis and septic shock [5,6].

Ischemia-modified albumin (IMA) is an albumin with a decreased affinity for cobalt ions resulting from ischemia and the production of reactive oxygen species. [8] The ischemia and oxygen free radicals lead to modification of the N-terminal sequence of albumin, which results in decreased affinity for metals, including cobalt ions [9,10]. Furthermore, ischemia and oxygen-free radicals lead to the breakdown of tissues, plasma phospholipids, and triglycerides, resulting in increased plasma-free fatty acid concentrations. The binding of plasma-free fatty acids to albumin leads to a decreased affinity for cobalt ions [11]. The IMA increases within 10 min after the ischemic insult, remains elevated for 6–12 h, and normalizes after 12–24 h [12,13]. Therefore, the IMA can be used as a biomarker to evaluate ischemic insults. Although few studies have evaluated the prognostic value of IMA in patients with sepsis or severe sepsis, they only included a small number of patients with septic shock and did not evaluate the prognostic value of the combination of IMA and lactate levels [14,15].

In this study, we aimed to evaluate the association between IMA and mortality in a larger population of patients with septic shock. Furthermore, we evaluated the prognostic value of the combination of IMA and lactate levels in predicting mortality in patients with septic shock in the emergency department.

2. Materials and Methods

2.1. Study Design and Setting

This retrospective observational study was conducted at Korea University Ansan Hospital, the only tertiary academic teaching hospital in Ansan-si, where 700,000 residents live. Approximately 50,000 patients visit the emergency department of Korea University Ansan Hospital annually. This study was conducted in accordance with the principles of the Declaration of Helsinki. This study was approved by the Institutional Review Board of Korea University Ansan Hospital (2024AS0039). The requirement for informed consent was waived by the Institutional Review Board because of the observational design of this study.

2.2. Study Population

Adults (age $\geq$ 18 years) patients diagnosed with septic shock in the emergency department from October 2019 to December 2021 were included in this study. Patients with do-not-resuscitate (DNR) orders, unknown 28-day mortality, and missing IMA results were excluded. Patients with acute ischemic and embolic events and terminal malignancy were excluded because acute ischemic insults in acute myocardial infarction, ischemic stroke, embolism, massive hemorrhage, and chronic ischemic states in terminal malignancy could influence the IMA [14,15].

2.3. Definitions and Data Collection

Sepsis was defined as an acute increase in the total sequential organ failure assessment (SOFA) score $\geq$ 2 from baseline due to infection [1]. Septic shock was defined as a serum lactate level > 2 mmol/L and the requirement of vasopressors despite adequate fluid resuscitation to maintain a mean arterial pressure $\geq$ 65 mmHg. All patients were managed according to the Surviving Sepsis Campaign guidelines [3].

The following patient data were collected from the electronic medical records: age, sex, comorbidities, age-adjusted Charlson Comorbidity Index (CCI) [16], SOFA score, infection focus, initial vital signs, initial laboratory results, and outcomes.

2.4. Measurements

Peripheral venous blood was obtained before initial resuscitation using plastic vacutainer serum-separating tubes (Belliver Industrial Estate, Plymouth, UK). After separation of serum by centrifuging at 3500 rpm for 10 min, IMA was quantified using an albumin-cobalt binding assay (Medicalsystem Biotechnology Co., Ltd., Ningbo, China) with a Cobas c702 module analyzer (Roche Diagnostics GmbH, Mannheim, Germany), following the manufacturer's instructions [15].

2.5. Outcome

The primary outcome was 28-day mortality. The secondary outcome was 90-day mortality.

2.6. Sample Size Calculations

Based on a subgroup analysis of 53 patients with septic shock in a previous study [15], the event rate was 26.4%, the adjusted hazard ratio (aHR) of IMA was 1.19, and the area under the receiver operating characteristic curve (AUROC) was 0.786. With a power of 0.9 and a significance level of 0.05, at least 63 patients are needed to evaluate the association between IMA and mortality in patients with septic shock, and at least 54 patients are needed to evaluate the prognostic value of IMA in patients with septic shock. However, to enhance the robustness and generalizability of our findings, we aimed to include a larger population of patients with septic shock.

2.7. Statistical Analysis

Continuous variables with normal distribution are expressed as means ± standard deviations and compared using the Student's *t*-test. Continuous variables without normal distributions are expressed as medians [interquartile range] and compared using the Mann–Whitney U test. Categorical variables are expressed as numbers (percentages) and compared using the chi-square test or Fisher's exact test.

The correlations between IMA and lactate, and IMA and SOFA scores were evaluated using Pearson's correlation coefficient.

To evaluate the independent association between IMA and outcomes, a multivariable Cox proportional hazards model was used. Variables with a *p*-value < 0.1 in the univariable Cox proportional hazard model (Table S1) were entered into the multivariable Cox proportional hazard model.

Receiver operating characteristic (ROC) curves of the IMA, lactate, and combinations of IMA and lactate for outcomes were obtained. The AUROC was compared using the DeLong test. The optimal cut-off point was determined using the Youden index.

According to the optimal cutoff points of IMA and lactate, the study population was categorized into high lactate/high IMA, high lactate/low IMA, low lactate/high IMA, and low lactate/low IMA groups. Kaplan–Meier curves and log-rank tests were used to compare the outcomes between the groups. A multivariable Cox proportional hazards model was used to evaluate independent associations between the groups and outcomes.

Subgroup analysis according to infection focus (respiratory and non-respiratory origin) was performed.

Sensitivity analysis was performed after multiple imputations using Multivariate Imputation through Chained Equations ('mice' package) for patients with missing IMA results [17].

Statistical significance was set at $p < 0.05$. Bonferroni corrections were used in the post-hoc analysis. All statistical analyses were performed using R version 4.0.2 (R Foundation for Statistical Computing, Vienna, Austria).

3. Results

We screened the data of 433 adults with septic shock between October 2019 and December 2021. Of these, 47 patients with DNR orders, 21 patients with unknown 28-day mortality, 11 patients with acute ischemic or embolic events (one acute myocardial infarction, one acute cerebral infarction, two pulmonary thromboembolisms, and seven severe gastrointestinal bleeding), 3 patients with terminal malignancy, and 57 patients with missing IMA results were excluded. Ultimately, 294 patients were included in the analysis (Figure 1). Mean age was 71.2 ± 14.1 years, 170 (57.8%) were men, and the mean SOFA score was 9.5 ± 2.7. The 28-day mortality rate was 31.3% (92/294) and the 90-day mortality rate was 41.0% (116/283).

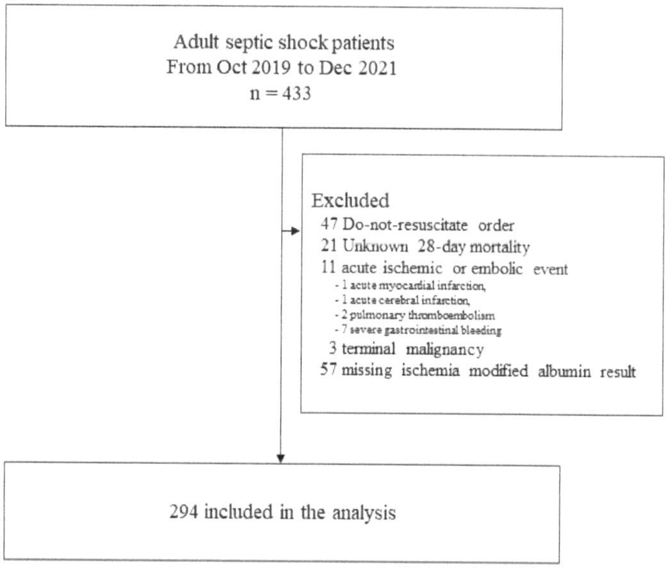

Figure 1. Flow chart of the study population.

3.1. Baseline Characteristics

The baseline characteristics according to 28-day mortality are shown in Table 1. The initial SOFA score and age-adjusted CCI were significantly higher in the non-surviving group than in the surviving group. The infection focus of the respiratory origin was more frequent in the non-surviving group than in the surviving group. Body temperature and hemoglobin and albumin levels were significantly lower in the non-surviving group than in the surviving group. Lactate (9.0 [4.4–14.8] vs. 3.8 [2.9–6.1] mmol/L, $p < 0.001$) and IMA (89.1 ± 7.2 vs. 83.8 ± 6.2 U/mL, $p < 0.001$) were significantly higher in the non-surviving group than the surviving group. The baseline characteristics according to 90-day mortality (n = 283) were similar (Table S2).

Table 1. Baseline characteristics according to 28-day mortality.

Variables	Survived on Day 28 (n = 202)	Died (n = 92)	p-Value
Sex			0.558
Men	114 (56.4%)	56 (60.9%)	
Women	88 (43.6%)	36 (39.1%)	
Age (years)	74 [62–81]	76 [64–83]	0.378
Infection focus			<0.001
Respiratory	58 (28.7%)	52 (56.5%)	
Gastrointestinal	28 (13.9%)	10 (10.9%)	
Biliary	45 (22.3%)	7 (7.6%)	
Genitourinary	58 (28.7%)	15 (16.3%)	
Others	13 (6.4%)	8 (8.7%)	
SOFA score	9 [7–10]	11 [9–12]	<0.001
Comorbidities			
Age-adjusted Charlson Comorbidity Index	4.4 ± 2.3	5.0 ± 2.6	0.050
Diabetes mellitus	80 (39.6%)	39 (42.4%)	0.746
Hypertension	123 (60.9%)	48 (52.2%)	0.201
Cardiac disease	29 (14.4%)	8 (8.7%)	0.243
Liver disease	9 (4.5%)	6 (6.5%)	0.645
Chronic kidney disease	19 (9.4%)	14 (15.2%)	0.206
Lung disease	7 (3.5%)	5 (5.4%)	0.636
Stroke	32 (15.8%)	12 (13.0%)	0.655
Malignancy	39 (19.3%)	25 (27.2%)	0.173

Table 1. *Cont.*

Variables	Survived on Day 28 (n = 202)	Died (n = 92)	p-Value
Initial vital signs			
Systolic blood pressure (mmHg)	106 [91–123]	102 [87.5–130.5]	0.432
Diastolic blood pressure (mmHg)	63 [55–75]	61 [52–76]	0.339
Heart rate (/min)	104 [90–122]	110.5 [88–128]	0.231
Respiratory rate (/min)	20 [18–23]	22 [16–28]	0.471
Body temperature	37.3 ± 1.3	36.6 ± 1.3	<0.001
Laboratory results			
Hemoglobin (g/dL)	11.9 ± 2.6	11.1 ± 3.0	0.034
White blood cell (*10^3/μL)	12.0 ± 8.4	10.6 ± 8.8	0.178
Platelets (*10^3/μL)	156 [108–233]	168 [82–253]	0.872
Creatinine (mg/dL)	1.4 [0.9–2.0]	1.6 [1.0–2.4]	0.194
Total bilirubin (mg/dL)	0.8 [0.5–1.4]	0.9 [0.4–1.8]	0.357
CRP (mg/dL)	9.4 [3.2–20.8]	10.8 [6.7–19.8]	0.303
Procalcitonin (ng/mL)	6.4 [1.2–26.0]	4.2 [1.4–17.0]	0.415
Lactate (mmol/L)	3.8 [2.9–6.1]	9.0 [4.4–14.8]	<0.001
Albumin (g/dL)	3.5 [3.1–3.9]	3.0 [2.6–3.3]	<0.001
Ischemia-modified albumin (U/mL)	83.8 ± 6.2	89.1 ± 7.2	<0.001

Data are presented as median [interquartile range], mean ± standard deviation, or number (%), as appropriate. Abbreviations: SOFA, sequential organ failure assessment; CRP, C-reactive protein.

3.2. Correlation Test

IMA was not correlated with lactate (R = 0.059, p = 0.314). IMA was not correlated with the SOFA score (R = 0.055, p = 0.348).

3.3. Multivariable Cox Proportional Hazards Model

IMA was independently associated with 28-day mortality after adjusting for infection focus, SOFA score, age-adjusted CCI, body temperature, hemoglobin, lactate, and albumin levels (aHR: 1.075, 95% confidence interval [CI]: 1.016–1.138, p = 0.012; Table 2). IMA was independently associated with 90-day mortality after adjustment (aHR 1.057, 95% CI 1.003–1.113, p = 0.038; Table S3).

Table 2. Multivariable Cox proportional hazards model for 28-day mortality.

Variables	Adjusted Hazard Ratio	95% Confidence Interval	p-Value
Infection focus			
Respiratory	Reference		
Gastrointestinal	0.789	0.391–1.591	0.507
Biliary	0.330	0.147–0.742	0.007
Genitourinary	0.507	0.277–0.929	0.028
Others	0.721	0.328–1.583	0.414
SOFA score	1.108	1.009–1.216	0.031
Age-adjusted Charlson Comorbidity Index	1.116	1.018–1.224	0.019
Body temperature	0.990	0.830–1.180	0.907
Hemoglobin (g/dL)	1.055	0.966–1.152	0.233
Lactate (mmol/L)	1.104	1.071–1.139	<0.001
Albumin (g/dL)	0.845	0.412–1.734	0.646
Ischemia-modified albumin (U/mL)	1.075	1.016–1.138	0.012

Abbreviations: SOFA, sequential organ failure assessment.

The formula of the multivariable Cox proportional hazards model is shown below.

$$h(t) = h_0(t) \times \exp(\beta_1 \times \text{Infection focus} + 1.108 \times \text{SOFA score} + 1.116 \times \text{age-adjusted CCI} + 0.990 \times \text{Body temperature} + 1.055 \times \text{Hemoglobin} + 1.104 \times \text{Lactate} + 0.845 \times \text{Albumin} + 1.075 \times \text{IMA}) \quad (1)$$

β_1 for Infection focus: 0.789 (gastrointestinal infection), 0.330 (biliary infection), 0.507 (genitourinary infection), and 0.721 (others)

3.4. Receiver Operating Characteristic Curve

The AUROC of IMA for 28-day mortality was 0.712 (95% CI: 0.648–0.775, $p < 0.001$; Figure 2A and Table S4). The optimal cutoff of IMA was 89.1 U/mL with a sensitivity of 53.3% and a specificity of 83.2%.

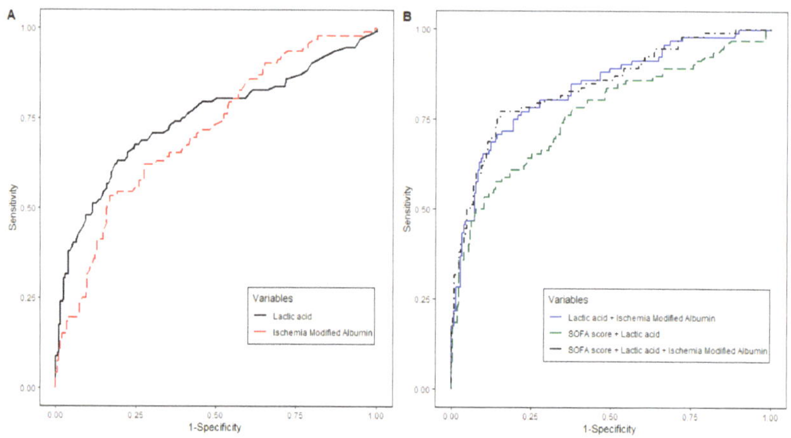

Figure 2. Receiver operating characteristic curve for 28-day mortality: (**A**) The black line shows the receiver operating characteristic curve of lactate and the red long-dashed line shows the receiver operating characteristic curve of IMA; (**B**) The blue line shows the receiver operating characteristic curve of a combination of lactate and IMA, the green long-dashed line shows the receiver operating characteristic curve of combination of SOFA score and lactate, and the black dot-dashed line shows the receiver operating characteristic curve of the combination of SOFA score, lactate, and IMA. Abbreviations: IMA, ischemia-modified albumin; SOFA, sequential organ failure assessment.

The AUROC of lactate for 28-day mortality was 0.746 (95% CI: 0.679–0.813, $p < 0.001$; Figure 2A and Table S4). The optimal cutoff of lactate was 7.3 mmol/L with a sensitivity of 63.0% and a specificity of 80.7%. The AUROC was comparable between the IMA and lactate levels ($p = 0.481$; Table S5).

The AUROC of the combination of IMA and lactate levels for 28-day mortality was 0.838 (95% CI: 0.786–0.889, $p < 0.001$; Figure 2B and Table S4). The AUROC of the combination of IMA and lactate was comparable to that of the combination of IMA, lactate, and SOFA score ($p = 0.576$). The AUROC of the combination of IMA and lactate was significantly higher than that of lactate ($p < 0.001$), IMA ($p < 0.001$), or the combination of lactate and SOFA scores ($p = 0.004$; Table S5).

3.5. Analysis According to Group Categorized by Optimal Cutoffs of Lactate and Ischemia-Modified Albumin

The 28-day mortality rate was significantly higher in the high lactate/high IMA group than in the other groups (high lactate/high IMA, high lactate/low IMA, low lactate/high IMA, and low lactate/low IMA groups: 86.1% vs. 44.3% vs. 38.3% vs. 10.7%, respectively, $p < 0.001$; Table 3). The 28-day mortality was significantly different between the groups, except for the difference between the high lactate/low IMA and low lactate/high IMA groups in the post-hoc analysis (Table S6).

Table 3. The 28-day mortality according to groups by optimal cutoffs.

	High Lactate (n = 97)	Low Lactate (n = 197)
High ischemia-modified albumin (n = 83)	31/36 (86.1%)	18/47 (38.3%)
Low ischemia-modified albumin (n = 211)	27/61 (44.3%)	16/150 (10.7%)

The Kaplan–Meier curve showed a significantly higher 28-day mortality in the high lactate/high IMA group than in the other groups (log-rank test, $p < 0.001$; Figure 3). The post-hoc analysis showed similar results.

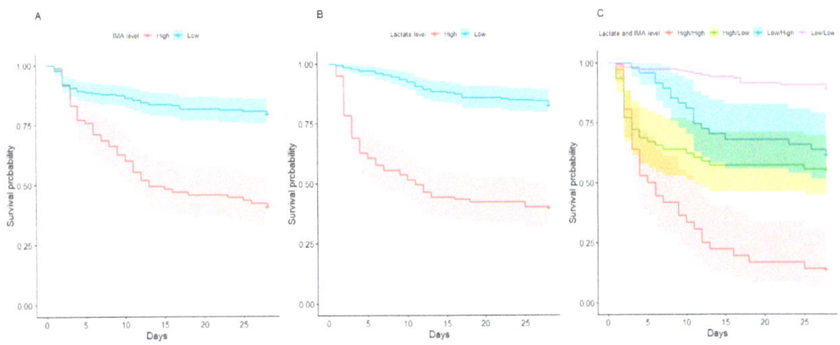

Figure 3. Kaplan–Meier curve according to groups: (**A**) Kaplan–Meier curve according to IMA level. The red line shows the survival probability of the high IMA group and the blue line shows the survival probability of the low IMA group; (**B**) Kaplan–Meier curve according to lactate level. The red line shows the survival probability of the high lactate group and the blue line shows the survival probability of the low lactate group; (**C**) Kaplan–Meier curve according to lactate and IMA level. The red line shows the survival probability of the high lactate/high IMA group, the green line shows the survival probability of the high lactate/low IMA group, the blue line shows the survival probability of the low lactate/high IMA group, and the purple line shows survival probability of low lactate/low IMA group. Abbreviations: IMA, ischemia modified albumin.

In the multivariable Cox proportional hazards model, the high lactate/high IMA, high lactate/low IMA, and low lactate/high IMA groups were independently associated with higher 28-day mortality than the low lactate/low IMA group after adjustments (high lactate/high IMA group: aHR 8.956, 95% CI: 4.071–19.70, $p < 0.001$; high lactate/low IMA group: aHR 4.510, 95% CI: 2.357–8.630, $p < 0.001$; low lactate/high IMA group: aHR 2.757, 95% CI: 1.214–6.262, $p = 0.015$; Table S7).

3.6. Subgroup Analysis

In the subgroup analysis, according to infection focus (respiratory and non-respiratory origin), IMA was independently associated with 28-day mortality in both subgroups (aHR 1.089, 95% CI: 1.039–1.141, $p < 0.001$ in subgroup with infection focus of respiratory origin and aHR 1.133, 95% CI: 1.032–1.243, $p = 0.009$ in subgroup with infection focus of non-respiratory origin).

3.7. Sensitivity Analysis

In the sensitivity analysis after multiple imputations for missing IMA results, IMA was independently associated with 28-day mortality after adjusting for the aforementioned covariables (aHR 1.083, 95% CI: 1.051–1.117, $p < 0.001$).

4. Discussion

In patients with septic shock, IMA was independently associated with mortality. The AUROC of IMA was comparable to that of lactate levels. The combination of lactate and IMA resulted in a better prediction of mortality than lactate alone, IMA alone, or a combination of the SOFA score and lactate. When patients with septic shock were categorized according to the optimal cut-offs of lactate and IMA, the high lactate/high IMA group showed an extremely high risk of mortality, the high lactate/low IMA and low lactate/high IMA groups showed an intermediate risk of mortality, and the low lactate/low IMA group showed a low risk of mortality.

Previous studies that evaluated the prognostic value of IMA in patients with sepsis or severe sepsis reported an independent association between IMA and short-term mortality and a similar AUROC of IMA for short-term mortality, which supports our results [14,15]. However, these previous studies included only 117 and 124 patients with sepsis; of these, 34 and 53 patients with septic shock were included, respectively [14,15]. Another previous study reported the poor prognostic value of IMA; however, this study included a small number of patients with sepsis and did not adjust for important covariables such as lactate level, SOFA score, or comorbidities [18]. Our study included a large number of patients with septic shock (n = 294), leading to more generalized results, particularly for patients with septic shock. Therefore, the IMA can be a good prognostic marker for the early prediction of mortality in patients with septic shock.

The Kaplan–Meier curve according to lactate and IMA levels showed different patterns of increased mortality in specific periods. The high-lactate group showed a sharp decline in survival probability within a few days of septic shock, whereas the high-IMA group showed a gradual decline in survival probability for up to 14 days after septic shock. This pattern of the Kaplan–Meier curve of IMA was similar to that reported in previous studies [14,15]. When patients were categorized according to lactate and IMA levels, high lactate and IMA patterns remained similar in the subgroups.

The difference in the survival pattern, the result showing no correlation between IMA and lactate, and the independent association of IMA after adjustments, including lactate, can be explained by the different mechanisms of lactate and IMA production. Inadequate oxygen delivery occurs in patients with septic shock [19], resulting in anaerobic metabolism that produces lactate. Furthermore, mitochondrial dysfunction in sepsis leads to increased anaerobic metabolism, despite adequate oxygen delivery [7]. As the time from inadequate oxygen delivery to death is short [7], the high lactate group had a high risk of death within a few days of septic shock.

Sepsis causes an imbalance between reactive oxygen species and antioxidants [20]. Alterations in the endothelium, neutrophils, macrophages, and mitochondrial signaling pathways during sepsis significantly induce oxidative stress [20]. Ischemic insult is associated with oxidative stress, and the production of reactive oxygen species affects albumin and produces IMA [8–11]. The high IMA group experienced high oxidative stress, which is associated with sepsis-induced endothelial dysfunction that causes microcirculatory failure and multiorgan failure [19]. This organ dysfunction may contribute to the continuous risk of death and lead to a gradual decline in survival probability for up to 14 days after septic shock. Therefore, in addition to lactate levels, IMA can be used to predict the prognosis of patients with septic shock.

Notably, patients with septic shock with both high lactate and high IMA showed an extremely high risk of mortality, whereas those with any single increase in lactate or IMA showed an intermediate risk of mortality, and those with both low lactate and low IMA showed an extremely low risk of mortality. These results were maintained even after adjusting for other variables. This might be due to the independent mechanisms of lactate and IMA production, as described above, which may have synergistic effects on predictive power. Furthermore, the combination of lactate and IMA showed good performance in predicting mortality in patients with septic shock, which is comparable to the combination of SOFA score, lactate, and IMA, and is better than the combination of the SOFA score and lactate. Therefore, any elevation of lactate or IMA and the combination of lactate and IMA would be helpful in the early risk stratification of patients with septic shock in the early stages of management. Moreover, physicians must be aware of the extremely high mortality risk of septic shock patients with both elevated lactate and IMA levels because significantly insufficient oxygen supply and substantial ischemic insult occur in these patients.

As the production of IMA is caused by ischemia and reactive oxygen species and is not infection-specific, IMA can be another representative marker of patient severity. IMA is associated with poor prognosis in critically ill diseases, such as aortic dissection [21,22], traumatic brain injury [23], and cardiac arrest [24,25]. It would be more helpful if lactate and IMA levels were used together for risk stratification in critically ill patients; however, this requires further study.

Although both the SOFA score and IMA were higher in the non-surviving group than in the surviving group, IMA was not correlated with the SOFA score. Additionally, IMA was independently associated with mortality even after adjusting for the SOFA score. This suggests that IMA could be an additional prognostic marker besides the SOFA score. However, this study only assessed SOFA scores on the first day of admission. Therefore, it remains unclear whether IMA is correlated with SOFA score after a few days of admission or with the worst SOFA score during the hospital stay. Further research is needed to explore these relationships.

The strength of our study is that we included a large number of patients with septic shock compared to previous studies [14,15,18]. Furthermore, we evaluated the correlation between IMA and lactate and the independent association of IMA with mortality after adjusting for important covariables, such as lactate, SOFA score, and comorbidities, and directly compared the performance of lactate and IMA in predicting mortality. In addition to 28-day mortality, we first evaluated 90-day mortality and found that IMA was associated with long-term mortality. Moreover, to the best of our knowledge, this is the first study to evaluate the usefulness of the combination of lactate and IMA levels in predicting the prognosis of patients with septic shock.

This study had some limitations. First, as this was an observational study, there could have been missed covariables. In addition, we could only find associations and not causalities. Second, this study was conducted at a single center. These results cannot be generalized to the entire population. Third, the IMA can differ according to the measurement protocols in each laboratory setting. The results and measurements must be validated before they can be applied. Fourth, some patients were excluded owing to missing IMA results. However, similar results were found in the sensitivity analysis after multiple imputations for missing IMA results.

5. Conclusions

In patients with septic shock, the elevation of ischemia-modified albumin was associated with mortality. The combination of ischemia-modified albumin and lactate showed good performance in predicting short-term mortality and can be a helpful tool for the early risk stratification of patients with septic shock.

Supplementary Materials: The following supporting information can be downloaded at: https://www.mdpi.com/article/10.3390/biomedicines12071421/s1, Table S1: Univariable Cox proportional hazard model; Table S2: Baseline characteristics according to 90-day mortality; Table S3: Multivariable Cox proportional hazard model for 90-day mortality; Table S4: Result of receiver operation characteristic curve for 28-day mortality; Table S5: Pairwise comparison of area under the receiver operating characteristic curves (AUROCs); Table S6: Pairwise comparison between groups; Table S7: Multivariable Cox proportional hazard model using groups according to optimal cutoffs.

Author Contributions: Conceptualization, B.-Y.J. and S.A.; methodology, B.-Y.J., J.-H.P., H.C., S.M. and S.A.; formal analysis, B.-Y.J., S.L., W.K. and S.A.; investigation, B.-Y.J., J.-H.P., H.C., S.M. and S.A.; data curation, B.-Y.J., S.L., W.K. and S.A.; writing—original draft preparation, B.-Y.J. and S.A.; writing—review and editing, B.-Y.J. and S.A.; visualization, B.-Y.J. and S.A.; supervision, S.A. All authors have read and agreed to the published version of the manuscript.

Funding: This work was supported by the Basic Science Research Program through the National Research Foundation of Korea (NRF), funded by the Ministry of Education (RS-2023-00272176). This funding source had no role in the design of this study and did not play any role during its execution, analysis, interpretation of the data, or decision to submit the results.

Institutional Review Board Statement: The Institutional Review Board approved this study (2024AS0039) and waived the requirement for informed consent owing to the nature of the study.

Informed Consent Statement: Patient consent was waived due to the nature of the study.

Data Availability Statement: The datasets used and/or analyzed in the current study are available from the corresponding author upon reasonable request.

Conflicts of Interest: The authors declare no conflicts of interest.

References

1. Singer, M.; Deutschman, C.S.; Seymour, C.W.; Shankar-Hari, M.; Annane, D.; Bauer, M.; Bellomo, R.; Bernard, G.R.; Chiche, J.-D.; Coopersmith, C.M.; et al. The third international consensus definitions for sepsis and septic shock (Sepsis-3). *JAMA* **2016**, *315*, 801–810. [CrossRef] [PubMed]
2. Reinhart, K.; Daniels, R.; Kissoon, N.; Machado, F.R.; Schachter, R.D.; Finfer, S. Recognizing sepsis as a global health priority—A WHO resolution. *N. Engl. J. Med.* **2017**, *377*, 414–417. [CrossRef] [PubMed]
3. Evans, L.; Rhodes, A.; Alhazzani, W.; Antonelli, M.; Coopersmith, C.M.; French, C.; Machado, F.R.; Mcintyre, L.; Ostermann, M.; Prescott, H.C.; et al. Executive summary: Surviving sepsis campaign: International guidelines for the management of sepsis and septic shock 2021. *Crit. Care Med.* **2021**, *49*, 1974–1982. [CrossRef] [PubMed]
4. Namgung, M.; Ahn, C.; Park, Y.; Kwak, I.-Y.; Lee, J.; Won, M. Mortality among adult patients with sepsis and septic shock in Korea: A systematic review and meta-analysis. *Clin. Exp. Emerg. Med.* **2023**, *10*, 157–171. [CrossRef] [PubMed]
5. Vincent, J.-L.; De Backer, D. Circulatory shock. *N. Engl. J. Med.* **2013**, *369*, 1726–1734. [CrossRef] [PubMed]
6. Vincent, J.-L.; Quintairos e Silva, A.; Couto, L.; Taccone, F.S. The value of blood lactate kinetics in critically ill patients: A systematic review. *Crit. Care* **2016**, *20*, 257. [CrossRef] [PubMed]
7. Suetrong, B.; Walley, K.R. Lactic acidosis in sepsis: It's not all anaerobic: Implications for diagnosis and management. *Chest* **2016**, *149*, 252–261. [CrossRef] [PubMed]
8. Shevtsova, A.; Gordiienko, I.; Tkachenko, V.; Ushakova, G. Ischemia-modified albumin: Origins and clinical implications. *Dis. Markers* **2021**, *2021*, 9945424. [CrossRef] [PubMed]
9. Bal, W.; Sokołowska, M.; Kurowska, E.; Faller, P. Binding of transition metal ions to albumin: Sites, affinities and rates. *Biochim. Biophys. Acta* **2013**, *1830*, 5444–5455. [CrossRef] [PubMed]
10. Watanabe, H.; Imafuku, T.; Otagiri, M.; Maruyama, T. Clinical implications associated with the posttranslational modification–induced functional impairment of albumin in oxidative stress–related diseases. *J. Pharm. Sci.* **2017**, *106*, 2195–2203. [CrossRef]
11. Lu, J.; Stewart, A.J.; Sadler, P.J.; Pinheiro, T.J.T.; Blindauer, C.A. Allosteric inhibition of cobalt binding to albumin by fatty acids: Implications for the detection of myocardial ischemia. *J. Med. Chem.* **2012**, *55*, 4425–4430. [CrossRef] [PubMed]
12. Sinha, M.K.; Vazquez, J.M.; Calvino, R.; Gaze, D.C.; Collinson, P.O.; Kaski, J.C. Effects of balloon occlusion during percutaneous coronary intervention on circulating Ischemia Modified Albumin and transmyocardial lactate extraction. *Heart* **2006**, *92*, 1852–1853. [CrossRef] [PubMed]
13. Bar-Or, D.; Winkler, J.V.; VanBenthuysen, K.; Harris, L.; Lau, E.; Hetzel, F.W. Reduced albumin-cobalt binding with transient myocardial ischemia after elective percutaneous transluminal coronary angioplasty: A preliminary comparison to creatine kinase-MB, myoglobin, and troponin I. *Am. Heart J.* **2001**, *141*, 985–991. [CrossRef] [PubMed]
14. Yin, M.; Liu, X.; Chen, X.; Li, C.; Qin, W.; Han, H.; Guo, H.; Yang, H.; Cao, D.; Du, Z.; et al. Ischemia-modified albumin is a predictor of short-term mortality in patients with severe sepsis. *J. Crit. Care* **2017**, *37*, 7–12. [CrossRef] [PubMed]
15. Park, J.; Ahn, S.; Lee, S.; Song, J.; Moon, S.; Kim, J.; Cho, H. Association of ischemia modified albumin with mortality in qSOFA positive sepsis patients by sepsis-3 in the emergency department. *Am. J. Emerg. Med.* **2021**, *44*, 72–77. [CrossRef] [PubMed]
16. Charlson, M.E.; Pompei, P.; Ales, K.L.; MacKenzie, C.R. A new method of classifying prognostic comorbidity in longitudinal studies: Development and validation. *J. Chronic Dis.* **1987**, *40*, 373–383. [CrossRef] [PubMed]
17. Van Buuren, S.; Groothuis-Oudshoorn, K. mice: Multivariate Imputation by Chained Equations in R. *J. Stat. Softw.* **2011**, *45*, 1–67. [CrossRef]
18. Cetin, M.; Oray, N.; Bayram, B.; Calan, O. The prognostic value of ischemia-modified albumin in patients with sepsis. *Niger. J. Clin. Pr.* **2021**, *24*, 680–684. [CrossRef] [PubMed]

19. Lelubre, C.; Vincent, J.-L. Mechanisms and treatment of organ failure in sepsis. *Nat. Rev. Nephrol.* **2018**, *14*, 417–427. [CrossRef] [PubMed]
20. Joffre, J.; Hellman, J. Oxidative stress and endothelial dysfunction in sepsis and acute inflammation. *Antioxid. Redox Signal.* **2021**, *35*, 1291–1307. [CrossRef] [PubMed]
21. Yang, G.; Zhou, Y.; He, H.; Pan, X.; Chai, X. Ischemia-modified albumin, a novel predictive marker of in-hospital mortality in acute aortic dissection patients. *Front. Physiol.* **2019**, *10*, 1253. [CrossRef] [PubMed]
22. Tomandlova, M.; Novotny, T.; Staffa, R.; Smutna, J.; Krivka, T.; Kruzliak, P.; Slaby, O.; Kubicek, L.; Vlachovsky, R.; Radova, L.; et al. Kinetics of d-lactate and ischemia-modified albumin after abdominal aortic surgery and their ability to predict intestinal ischemia. *Clin. Biochem.* **2023**, *112*, 43–47. [CrossRef] [PubMed]
23. Radwan, T.A.; Fahmy, R.S.; El Emady, M.F.; Khedr, A.S.; Osman, S.H.; ElSonbaty, M.I.; El-Kholy, B.M.B.; Thabit, M.A.; Elkatatny, A.M. Ischemia-modified albumin as a biomarker for prediction of poor outcome in patients with traumatic brain injury: An observational cohort study. *J. Neurosurg. Anesthesiol.* **2019**, *33*, 254–257. [CrossRef] [PubMed]
24. Turedi, S.; Gunduz, A.; Mentese, A.; Dasdibi, B.; Karahan, S.C.; Sahin, A.; Tuten, G.; Kopuz, M.; Alver, A. Investigation of the possibility of using ischemia-modified albumin as a novel and early prognostic marker in cardiac arrest patients after cardiopulmonary resuscitation. *Resuscitation* **2009**, *80*, 994–999. [CrossRef] [PubMed]
25. Yucel, H.; Turkdogan, K.A.; Zorlu, A.; Aydin, H.; Kurt, R.; Yilmaz, M.B. Association between oxidative stress index and post-CPR early mortality in cardiac arrest patients: A prospective observational study. *Anatol. J. Cardiol.* **2015**, *15*, 737–743. [CrossRef] [PubMed]

biomedicines

MDPI

Article

Association of LVV-Hemorphin-7 with Sepsis and Shock: Roles of Cathepsin D and G in Hemoglobin Metabolism in a Prospective ICU Cohort Study

Yao-Kuang Wu [1,2,†], Hsueh-Wen Chung [3], Yi-Ting Chen [2,4,†], Hsing-Chun Chen [5], I-Hung Chen [5] and Wen-Lin Su [1,2,*]

1 Division of Pulmonary and Critical Care Medicine, Department of Internal Medicine, Taipei Tzu Chi Hospital, Buddhist Tzu Chi Medical Foundation, New Taipei 231, Taiwan; drbfci@gmail.com
2 School of Medicine, Tzu Chi University, Hualien 970, Taiwan; kateytc@gmail.com
3 Department of Nursing, College of Nursing, National Yang Ming Chiao Tung University, Taipei City 112, Taiwan; snow721103@gmail.com
4 Division of Pulmonary and Critical Care Medicine, Department of Internal Medicine, Hualien Tzu Chi Hospital, Buddhist Tzu Chi Medical Foundation, Hualien 970, Taiwan
5 Division of Pulmonary and Critical Care Medicine, Department of Internal Medicine, Dalin Tzu Chi Hospital, Buddhist Tzu Chi Medical Foundation, Chiayi County 622, Taiwan; dm689688@tzuchi.com.tw (H.-C.C.); b89401098@ntu.edu.tw (I.-H.C.)
* Correspondence: williamsu2007@gmail.com; Tel.: +886-2-66289779; Fax: +886-2-66289009
† These authors contributed equally to this work.

Abstract: Background: Sepsis is a leading cause of mortality in intensive care units (ICUs). Cell-free hemoglobin (CFH) released during sepsis interacts with lysosomal enzymes from neutrophils and macrophages. This study aims to examine the association of LVV-hemorphin-7 (LVV-H7), cathepsin D, and cathepsin G with sepsis and shock in ICU patients. Methods: A prospective observational cohort study was conducted in the medical ICU of a tertiary referral hospital in Taiwan. The patients with an acute increasing sequential organ failure assessment (SOFA) score ≥ 2 between 2022 and 2023. Blood samples from 40 healthy controls were obtained from the hospital biobank. CFH metabolites, including LVV-H7 and lysosomal enzyme cathepsin D and cathepsin G, were compared between the sepsis (definite and probable) and non-sepsis (possible sepsis) groups. Multivariate logistic regression analyzed factors associated with sepsis and shock. Results: Among 120 patients, 75 were classified as septic and 45 as non-septic. Significant differences were observed in CFH, cathepsin D, cathepsin G, and LVV-H7 levels between sepsis and non-sepsis groups. LVV-H7 was a significant predictor for sepsis (adjusted OR [aOR] 1.009, 95% CI 1.005–1.013; $p < 0.001$) and shock (aOR 1.005, 95% CI 1.002–1.008; $p < 0.05$). Cathepsin G predicted non-shock (aOR 0.917, 95% CI 0.848–0.991; $p < 0.05$), while cathepsin D predicted septic shock (aOR 1.001, 95% CI 1.000–1.002; $p < 0.05$). Conclusions: LVV-H7, cathepsin D, and cathepsin G are associated with the classification of sepsis and shock episodes in critically ill patients with elevated SOFA scores.

Keywords: sepsis; shock; cathepsin; LVV-hemorphin-7; sequential organ failure assessment score

Citation: Wu, Y.-K.; Chung, H.-W.; Chen, Y.-T.; Chen, H.-C.; Chen, I.-H.; Su, W.-L. Association of LVV-Hemorphin-7 with Sepsis and Shock: Roles of Cathepsin D and G in Hemoglobin Metabolism in a Prospective ICU Cohort Study. *Biomedicines* **2024**, *12*, 2789. https://doi.org/10.3390/biomedicines12122789

Academic Editor: Christian Lehmann

Received: 14 November 2024
Revised: 2 December 2024
Accepted: 5 December 2024
Published: 9 December 2024

1. Introduction

Sepsis is an extreme response to an infection that causes significant erythrocyte damage, particularly in intensive care units (ICUs) where patients often develop anemia [1–3]. During sepsis-induced anemia, cell-free hemoglobin (CFH) is released by erythrocytes through a series of mechanisms [4]. Traditionally, free heme and CFH-damaged tissue have been observed to worsen sepsis in lipopolysaccharide-induced systemic inflammatory animal models [5,6]. Clinical studies have shown that higher levels of CFH in patients with sepsis are associated with lower survival rates [7,8].

In infection cases, neutrophils and macrophages are recruited to the infection site to combat pathogen invasion. Cathepsin G originates from neutrophils [9,10], while cathepsin D comes from the lysosomes of macrophages [11]. These enzymes are potentially involved in mechanisms that play a critical role in Hb cleavage during sepsis and the production of hemorphins [12]. Hemorphin-7 (H7), which includes LVV-H7 and VV-H7, is defined as an Hb metabolite found in thrombotic tissue and abdominal aortic aneurysms that attracts leukocytes [13]. This suggests a synergistic effect between Hb and immune cells during sepsis and the possible release of LVV-H7 into the blood, attracting leucocyte recruitment. Recent cell culture studies have shown that LVV-H7 inhibits the angiotensin-converting enzyme, downregulating angiotensin II and decreasing blood pressure, suggesting a negative correlation between LVV-H7 levels and shock [14]. The LVV-H7 may have synergism that increases the risk of shock. These findings would justify the need for assessing the hemoglobin metabolism-related biomarkers of sepsis severity.

This study was the first investigation about CFH, lysosomal enzymes, and LVV-H7 in critical sepsis cases and compared these biomarkers between sepsis and control groups to identify potential diagnostic factors for sepsis and shock.

2. Methods

2.1. Study Design and Participant Enrolment

This prospective observational cohort study consisted of patients with suspected sepsis admitted to the medical intensive care units (ICUs) at a tertiary referral medical center in Taiwan from 2022 to 2023, as shown in Figure 1.

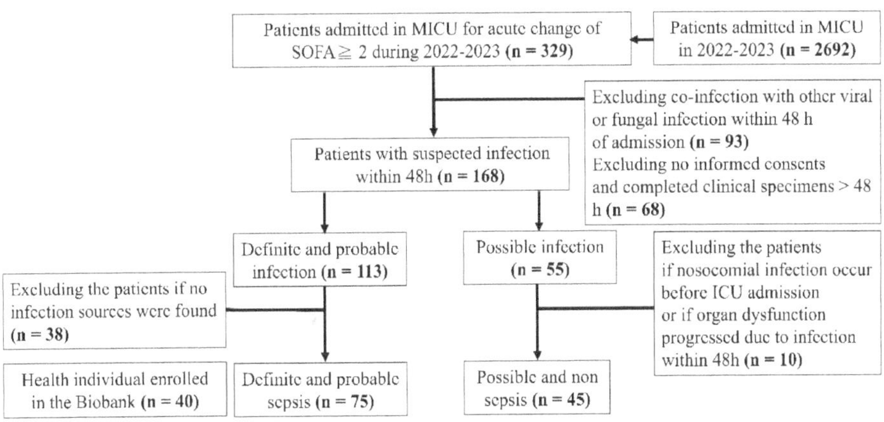

Figure 1. Cohort flow diagram of the sepsis study enrollment process. MICU: Medical intensive care unit; ICU: intensive care unit; SOFA, Sequential Organ Failure Assessment Score.

The study protocol titled "Exploring the anti-inflammation mechanisms of hemoglobin and its metabolites in sepsis" was approved by the Institutional Review Board of Taipei Tzu Chi Hospital, Buddhist Tzu Chi Medical Foundation, New Taipei City, Taiwan, on 16 March 2020 (Protocol No.: 08-P-134). To evaluate the Hb metabolite concentrations in the general population, the study titled "Analysis of erythrocyte-related metabolites in plasma of healthy adults using biobank specimens" was approved by the same board on 11 January 2024 (Protocol No.: 13-IRB006). Forty blood samples previously collected from healthy participants were applied as healthy controls. The informed consent was waived, and the Biobank Ethics Committee approved this application on 12 January 2024 (application No: Tzubiobank 2024-05). All study procedures adhered to the ethical standards of the responsible committee on human experimentation and the Helsinki Declaration of 1975. Informed consents were obtained from participants or their legal agents upon ICU admission.

Inclusion criteria included patients meeting the sequential organ failure assessment (SOFA) score ≥ 2 and suspected infection since initial ICU admission. Suspected infections were categorized as definite, probable, or possible based on clinical symptoms and radiological evidence according to the International Sepsis Forum Criteria [15]. The sepsis classifications were divided into definite, probable, and possible sepsis categories following the criteria established by Rhee et al. [16]. and Mellhammar et al. [17], as detailed in the following.

Definite and probable sepsis were defined by the presence of criterion 1 and either criterion 2 or 3:

1. A definite or probable source of infection (e.g., positive cultures or radiographic evidence with a compatible clinical syndrome).
2. Organ dysfunction due to infection without any discernable cause other than infection.
3. Organ dysfunction most likely attributable to infection, although other potential contributors were present.

Possible sepsis was defined by the presence of either criterion 1 or 2:

1. Patients were treated for presumed sepsis but lacked definitive evidence of infection.
2. Patients had alternative definite or possible explanations for organ dysfunction.

In our study, the sepsis group included patients classified as having definite or probable sepsis, while the non-sepsis group included those classified as having possible sepsis.

Exclusion criteria: Patients were excluded if they had co-infections with other viral or fungal pathogens within 48 h of admission or if the collection of clinical specimens was delayed beyond 48 h after admission, as these data would not reflect early sepsis conditions. After applying these exclusion criteria, patients were retrospectively classified into either the sepsis (definite and probable) or non-sepsis (possible sepsis) groups based on the established sepsis classification criteria.

Subgroup exclusion criteria: In the "definite and probable sepsis" group, patients were excluded if no source of microbiologic infection was identified. In the "possible and non-sepsis" group, patients were excluded if nosocomial infection occurred before ICU admission or if organ dysfunction progressed due to infection within 48 h.

The identified pathogens from cultures and the primary sources of infection in the sepsis (definite and probable) group were documented. The causes of organ dysfunction in the non-sepsis (possible sepsis) group were documented and are available in the following seven groups.

1. Acute or chronic renal failure leading to acute pulmonary edema with hypoxemic respiratory failure or hyperkalemia with bradycardic heart failure.
2. Cardiac dysfunction-related cardiogenic hypotension, including acute myocardial infarction, decompensated heart failure, and arrhythmias.
3. Pulmonary disease causing impaired gas exchange and hypoxemia, such as chronic obstructive pulmonary disease, asthma, and interstitial lung disease.
4. Gastrointestinal tract hemorrhage with hypovolemic hypotension, including esophageal varices bleeding and upper or lower gastrointestinal tract bleeding.
5. Acute pancreatitis with volume depletion and distributive hypotension.
6. Metabolic acidosis-related hypotension or cardiac suppression, including diabetic ketoacidosis and hyperglycemic hyperosmolar syndrome.
7. Anaphylaxis resulting in distributive hypotension.

Patients identified as having septic shock were those requiring vasopressors to maintain a mean arterial pressure (MAP) ≥ 65 mmHg [18] due to persistent hypotension with a serum lactate level > 2 mmol/L (18 mg/dL), despite sufficient volume resuscitation. All the patients adhered to the standard treatment protocol of the Surviving Sepsis Campaign guidelines [19].

2.2. Data Collection

Patient data, including age, sex, initial vital signs, quick SOFA, Glasgow Coma Scale (GCS), and Charlson Comorbidity Index (CCI) scores [20], were collected for demographic analysis. Regular laboratory blood tests were performed during enrollment. Sources of infection were documented through cultures of blood, tracheal aspiration, sputum, urine, body fluid, and pus. All bacteria were subjected to the minimum inhibitory concentration tests by using the VITEK®2 automated system (bioMérieux, Lyon, France). Antibiotic resistance was manually examined using the BBL Sensi-Disc test or automatically determined by the VITEK®2 automated system in our hospital. Additional tests for pneumonia pathogens included urine antigens of *Legionella pneumoniae* and *Streptococcus pneumoniae*, and immunoglobulin M of *Mycoplasma pneumoniae* and *Chlamydia pneumoniae*. In patients with sepsis, for the purpose of rapidly tracing multi-drug-resistant organisms, a film array (BioFire Diagnostics, Salt Lake City, UT, USA) was used as a pneumonia panel for specimens from tracheal aspiration and blood culture identification of bacterial growth in the blood culture panel at 48 h without definite culture results. Clinical outcomes, including the length of ICU stay, shock episodes within 48 h of ICU admission, hospitalization duration, and survival status, were recorded.

2.3. Free Hemoglobin, Cathepsin D, Cathepsin G, LVV-H7, and Angiotensin II Enzyme-Linked Immunosorbent Assay (ELISA) Analysis

Blood specimens from both ICU and healthy controls were stored at −80 °C until analysis. The angiotensin II ELISA (ADI-900-204; Enzo Life Sciences Inc., Farmingdale, NY, USA) was used, and all procedures were conducted according to the manufacturer's instructions. For Hb metabolism, LVV-H7 was selected as the final stable form. Subsequently, cathepsin D (ab119586; Abcam, Cambridge, UK), cathepsin G (EC3237-1, Assaypro, St. Charles, MO, USA), Hb (ab157707; Abcam, Cambridge, UK), and LVV-H7 (MBS8820153; MyBioSource, San Diego, CA, USA) antibodies were added to the wells. After incubation, the washing step was repeated, and streptavidin horseradish peroxidase solution was added. The plate was washed again, tetramethylbenzidine substrate was added, and the plate was incubated in the dark. A stop solution was used to terminate the reaction, and 450 nm was chosen as the reference wavelength to measure absorbance. All tests were performed in duplicate.

2.4. Statistical Analysis

Categorical data are presented as frequencies and percentages, and the continuous data are presented as mean ± standard deviation. A one-way analysis of variance was used to compare among the three groups. The two continuous variables were compared using Student's *t*-test, and categorical variables were compared using the chi-square test. Multivariate logistic regression analysis was performed, and confounding factors were adjusted using stepwise forward enrollment methods to identify Hb metabolite biomarkers for predicting sepsis, shock, and septic shock. Statistical analyses were performed using the IBM SPSS Statistics statistical software for Windows (version 26.0; IBM Corp, Armonk, NY, USA); statistical significance was set at $p < 0.05$.

3. Results

3.1. Hb Metabolic Biomarkers Among Sepsis, Non-Sepsis, and Healthy Control Groups

In total, 329 critically ill patients met the inclusion criteria of acute SOFA score change ≥ 2. After applying the exclusion/inclusion criterion, 120 patients were enrolled in the study (Figure 1), with 75 in the sepsis group and 45 in the non-sepsis groups. The control group consisted of blood samples from 40 healthy individuals. Post hoc analyses revealed that CFH levels were significantly higher ($p < 0.001$) in the sepsis group (15.3 ± 8.8 mg/dL) compared to the non-sepsis (9.8 ± 3.5 mg/dL) and control (9.3 ± 3.9 mg/dL) groups as Figure 2. Similarly, cathepsin D and LVV-H7 levels were significantly higher in the sepsis group (849 ± 703.1 and 371.7 ± 190.5 ng/mL, respectively) ($p < 0.001$) than in the non-sepsis (337.4 ± 141.1 and 194.5 ± 63.1 ng/mL, respectively) and control (220.4 ± 86.5 and 82.3 ± 6.4 ng/mL, respectively) groups. Cathepsin G levels were significantly higher ($p < 0.001$) in the sepsis (6.8 ± 7.0 µg/mL) and non-sepsis (4.5 ± 6.3 µg/mL) groups than in the control group (1.2 ± 0.3 µg/dL). However, angiotensin II levels were significantly lower in the sepsis (78.5 ± 105.8 ng/mL) and non-sepsis (87.8 ± 95.5 ng/mL) groups compared to the control group (527.1 ± 625.6 ng/mL). Significant differences in free Hb, cathepsin D, cathepsin G, LVV-H7, and angiotensin II values were observed between critically ill patients with acute changes in SOFA scores and the control group.

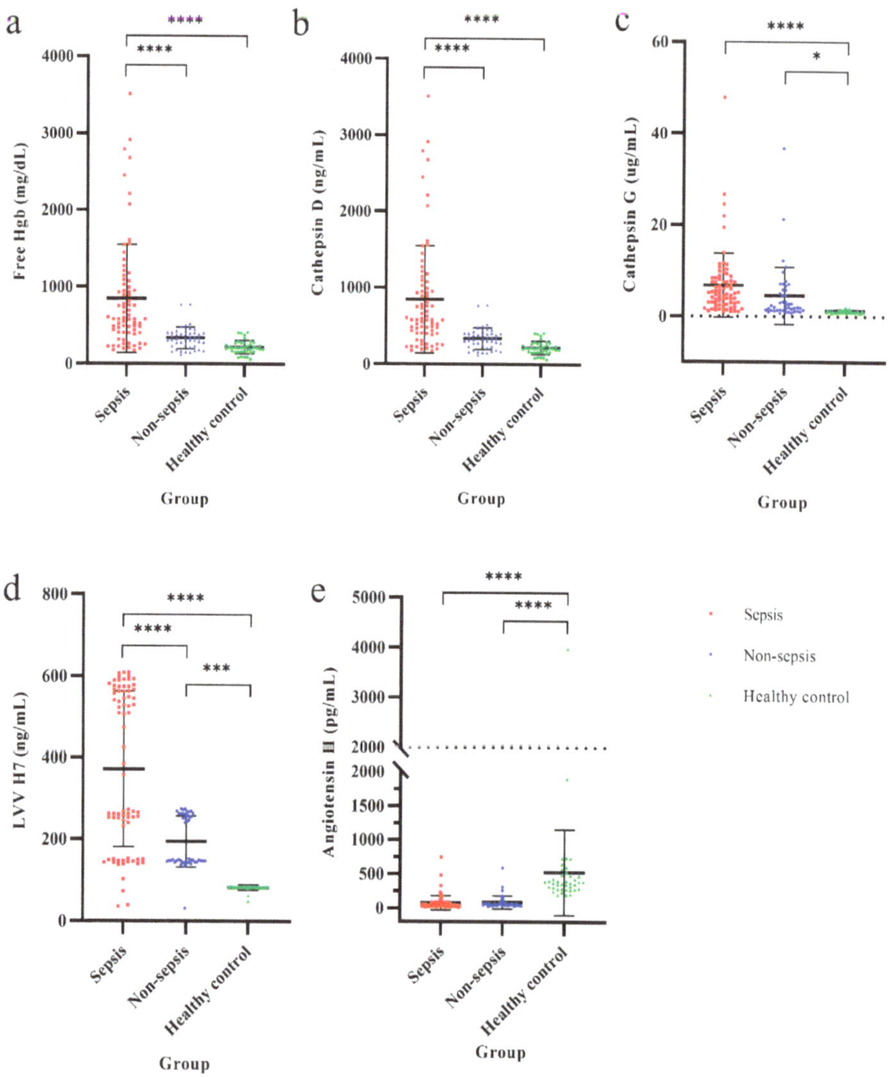

Figure 2. Hemoglobin metabolism among sepsis, non-sepsis, and health controls. (**a**) CFH, (**b**) cathepsin D, (**c**) cathepsin G, (**d**) LVV-H7 and (**e**) angiotensin II. CFH: Cell-free hemoglobin; LVV-H7: LVV-hemorphin-7. *p*-values calculated using Student's *t*-test are shown above the scatter points. * $p < 0.05$, *** $p < 0.001$, **** $p < 0.0001$ for the difference between paired scatter points, respectively.

3.2. *The Isolated Microorganisms in the Sepsis Group and the Diagnoses in the Non-Sepsis Group*

The distribution of pathogens and the primary sources of infection in the sepsis (definite and probable) group, along with the diagnoses contributing to organ dysfunction in the non-sepsis (possible sepsis) group, are presented in Figure 3.

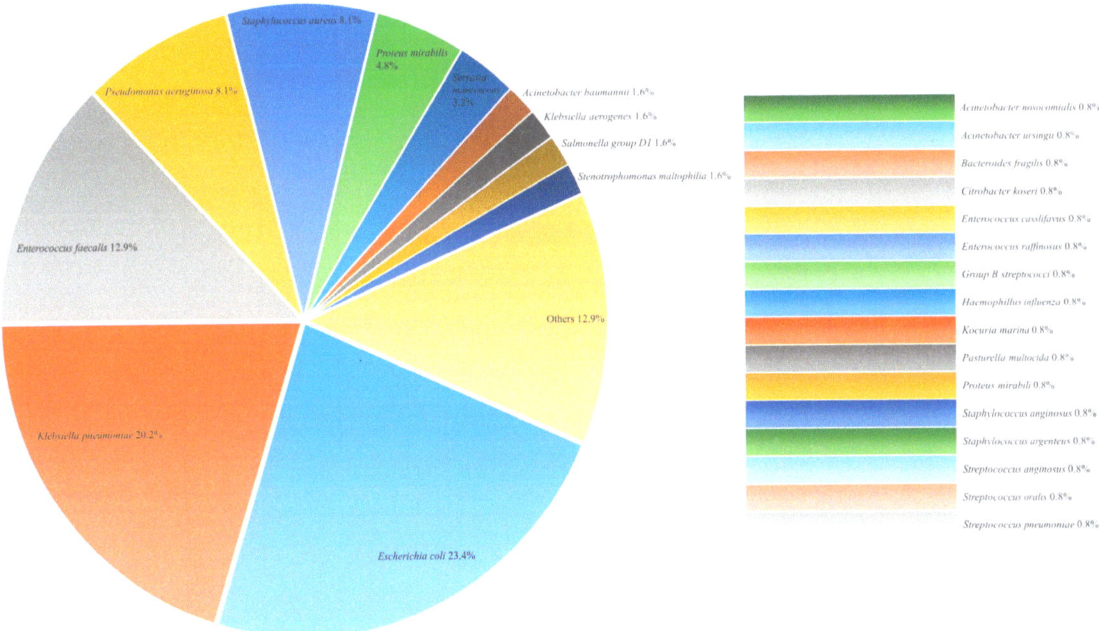

(A)

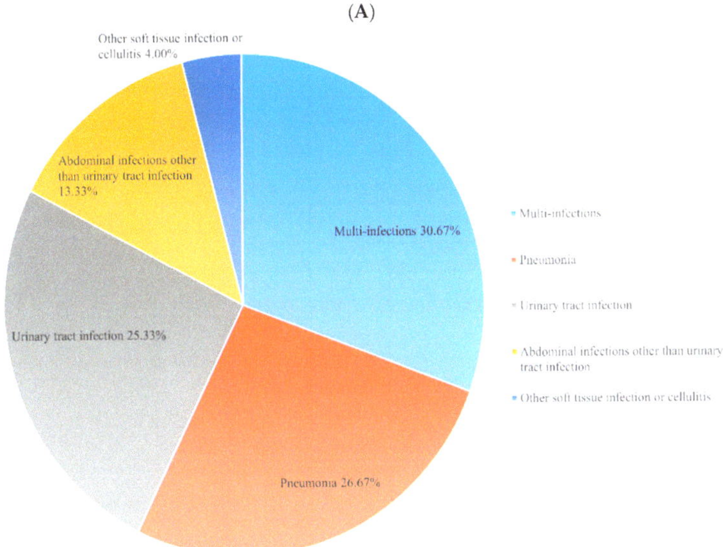

(B)

Figure 3. *Cont.*

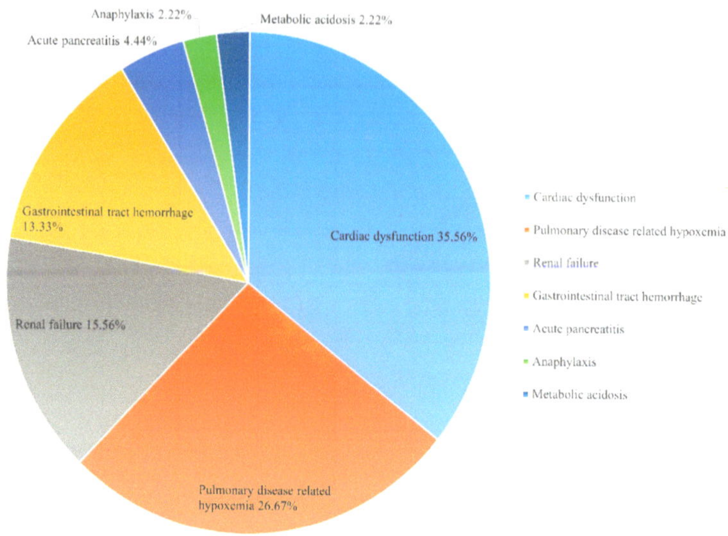

(C)

Figure 3. Etiologies of organ dysfunction in sepsis and non-sepsis groups. (**A**) Distribution of pathogens in the sepsis group; (**B**) Primary sources of infection in the sepsis group; (**C**) Major causes of multiple organ dysfunction in the non-sepsis group. The distribution of pathogens in the sepsis group was as follows: *Escherichia coli* (n = 29, 23.4%), *Klebsiella pneumoniae* (n = 25, 20.2%), *Enterococcus faecalis* (n = 16, 12.9%), *Pseudomonas aeruginosa* (n = 10, 8.1%), *Staphylococcus aureus* (n = 10, 8.1%), *Proteus mirabilis* (n = 6, 4.8%), *Serratia marcescens* (n = 4, 3.2%), and others (n = 24, 19.3%) (Figure 3). The primary sources of infection in the sepsis group included pneumonia (n = 20, 26.7%), urinary tract infection (n = 19, 25.3%), abdominal infections other than urinary tract infection (n = 10, 13.3%), other soft tissue infections or cellulitis (n = 3, 4.0%), and multi-infections (n = 23, 30.7%). The major causes of multiple organ dysfunction in the non-sepsis group were as follows: cardiac dysfunction (n = 16, 35.6%), pulmonary disease-related hypoxemia (n = 12, 26.7%), renal failure (n = 7, 15.6%), gastrointestinal tract hemorrhage (n = 6, 13.3%), acute pancreatitis (n = 2, 4.4%), metabolic acidosis (n = 1, 2.2%), and anaphylactic hypotension (n = 1, 2.2%).

3.3. Comparisons of Parameters Between Sepsis and Non Sepsis Groups

Table 1 shows the demographic characteristics of the sepsis and non-sepsis groups. Significant differences were observed in initial body temperature (BT; 36.99 vs. 36.38 °C, $p < 0.05$), MAP (82.5 vs. 97.6 mmHg, $p < 0.001$), and SOFA score (6.6 vs. 5.1, $p < 0.05$) between the sepsis and non-sepsis groups. Clinical variables and outcomes were similar between the two groups, except for shock episodes, which were significantly higher in the sepsis group (81.3% vs. 44.4%, $p < 0.001$).

There were no significant differences between the two groups in red blood cell count, Hb, platelets, monocytes, sodium (Na), potassium (K), alanine transaminase, aspartate transaminase, albumin, blood urea nitrogen, creatinine, random glucose, lactate, prothrombin time, international normalized ratio, pH, SaO_2 in arterial blood gas, ratio of arterial oxygen partial pressure to fractional inspired oxygen, cathepsin G, and angiotensin II levels (Table 2). However, significantly higher levels of white blood cell (WBC) count (13.16 vs. 9.36 $\times$ $10^3/\mu L$, $p < 0.001$), total bilirubin (1.71 vs. 0.95 mg/dL, $p < 0.05$), activated partial thromboplastin time (aPTT) (32.65 vs. 28.61 s, $p < 0.05$) levels, neutrophil (10.12 vs. 6.91 $\times$ $10^3/\mu L$, $p < 0.001$), C-reactive protein (CRP) (13.50 vs. 5.11 mg/dL, $p < 0.001$) were observed in the sepsis group than in the non-sepsis group, respectively. In contrast, lower lymphocyte levels (0.91 vs. 1.71 $\times$ $10^3/\mu L$, $p < 0.05$) were detected in the sepsis group.

Table 1. Characteristics of the different sepsis groups of the study population (n = 120).

Variables	Definite and Probable Sepsis (n = 75)	Possible and Non-Sepsis (n = 45)	p-Value
Age (years)	68.81 ± 14.17	71.09 ± 13.56	0.388 [c]
Sex			0.530 [a]
Female	31 (41.3%)	16 (35.6%)	
Male	44 (58.7%)	29 (64.4%)	
Body mass index (kg/m^2)	22.96 ± 4.73	22.85 ± 4.41	0.900 [c]
Vital signs			
Body temperature (°C)	36.99 ± 1.45	36.38 ± 0.81	0.004 [c*]
Respiratory rate (/min)	21.17 ± 5.38	20.76 ± 4.69	0.667 [c]
Heart rate (/min)	106.33 ± 25.76	100.16 ± 26.57	0.211 [c]
MAP (mmHg)	82.53 ± 22.00	97.58 ± 21.33	<0.001 [c**]
SpO_2 (%)	94.85 ± 4.94	94.36 ± 5.03	0.597 [c]
GCS	12.13 ± 4.06	12.27 ± 4.08	0.862 [c]
CCI	5.55 ± 2.80	5.38 ± 2.31	0.734 [c]
Comorbidities			
Diabetes	33 (44.0%)	18 (40.0%)	0.668 [a]
Cardiovascular disease	26 (34.7%)	22 (48.9%)	0.124[a]
Chronic kidney disease	24 (32.0%)	12 (26.7%)	0.537 [a]
Neurologic diseases	19 (25.3%)	11 (24.4%)	0.913 [a]
Pulmonary disease	15 (20.0%)	12 (26.7%)	0.397 [a]
Malignancy	15 (20.0%)	6 (13.3%)	0.352 [a]
Chronic liver disease	8 (10.7%)	4 (8.9%)	1.000 [b]
Autoimmune disease	2 (2.7%)	2 (4.4%)	0.630 [b]
Disease severity			
qSOFA	1.09 ± 0.83	0.84 ± 0.74	0.099 [c]
SOFA score	6.61 ± 2.91	5.11 ± 2.80	0.006 [c]
APACHE II score	23.07 ± 9.92	21.36 ± 8.40	0.336 [c]
Oxygenation device status			0.512 [a]
IMV	35 (46.7%)	26 (57.8%)	
NIV	5 (6.7%)	4 (8.9%)	
Oxygen supplement	31 (41.3%)	14 (31.1%)	
Oxygen not needed	4 (5.3%)	1 (2.2%)	
Clinical outcomes			
Length of stay in ICU	12.44 ± 12.84	15.31 ± 14.65	0.263 [c]
Hospital days	23.20 ± 19.71	27.36 ± 23.91	0.307 [c]
Shock episodes	61 (81.3%)	20 (44.4%)	<0.001 [a**]
Survival	54 (72.0%)	35 (79.5%)	0.360 [a]

[a] A chi-square test is used for the comparison of categorical variables, [b] Fisher's exact test is used for the comparison of categorical variables, [c] an independent t-test is used for continuous variables between the sepsis and non-sepsis groups. * $p < 0.05$; ** $p < 0.001$. APACHE, Acute Physiology and Chronic Health Evaluation; CCI, Charlson Comorbidity Index; GCS, Glasgow Coma Scale; MAP, mean arterial pressure; SOFA, Sequential Organ Failure Assessment; SpO_2, saturation from pulse oximeter; qSOFA, Quick Sequential Organ Failure Assessment; IMV, Invasive Mechanical Ventilation; NIV, Noninvasive Ventilation.

Table 2. Laboratory data from different groups of the study population (n = 120).

Variables	Definite and Probable Sepsis (n = 75)	Possible and Non-Sepsis (n = 45)	p-Value
WBC ($10^3/\mu L$)	13.16 ± 7.71	9.36 ± 4.27	0.001 *
RBC ($10^6/\mu L$)	3.67 ± 0.84	3.60 ± 0.92	0.650
Hemoglobin (g/dL)	11.00 ± 2.58	10.97 ± 2.58	0.948
Platelets ($10^3/\mu L$)	186.84 ± 129.37	202.96 ± 93.94	0.433
Neutrophil ($10^3/\mu L$)	10.12 ± 6.70	6.91 ± 3.64	0.001 *
Lymphocyte ($10^3/\mu L$)	0.91 ± 0.80	1.71 ± 1.96	0.011 *
Monocyte ($10^3/\mu L$)	0.50 ± 0.41	0.47 ± 0.31	0.602
Na (mEq/L)	135.13 ± 6.93	136.76 ± 5.22	0.149
K (mEq/L)	4.23 ± 1.04	4.17 ± 0.84	0.748

Table 2. *Cont.*

Variables	Definite and Probable Sepsis (n = 75)	Possible and Non-Sepsis (n = 45)	p-Value
AST (U/L)	92.60 ± 167.37	71.24 ± 147.65	0.481
ALT (U/L)	41.43 ± 59.41	39.49 ± 42.64	0.849
Albumin (g/dL)	3.00 ± 0.58	3.19 ± 0.49	0.075
BUN (mg/dL)	52.77 ± 41.87	43.53 ± 32.13	0.206
Creatinine (mg/dL)	2.93 ± 2.57	2.68 ± 3.19	0.642
Random Glucose (mg/dL)	179.30 ± 80.1	176.15 ± 82.68	0.837
Lactate (mmol/L)	3.62 ± 3.49	3.75 ± 4.57	0.863
CRP (mg/dL)	13.50 ± 11.38	5.11 ± 7.53	<0.001 **
PCT (ng/mL)	19.03 ± 37.18	14.57 ± 43.85	0.553
Total bilirubin (mg/dL)	1.71 ± 2.56	0.95 ± 0.77	0.019 *
PT (second)	12.30 ± 2.49	11.55 ± 1.84	0.084
INR (ratio)	1.23 ± 0.32	1.18 ± 0.32	0.404
aPTT (second)	32.65 ± 8.04	28.61 ± 6.12	0.002 *
ABG pH	7.36 ± 0.12	7.39 ± 0.10	0.093
ABG SaO_2	95.89 ± 3.43	96.82 ± 3.43	0.154
P/F ratio	243.97 ± 137.03	289.95 ± 148.78	0.088
Hemoglobin catabolism			
Free Hemoglobin (mg/dL)	15.27 ± 8.81	9.75 ± 3.51	<0.001 **
Cathepsin D (ng/mL)	849.19 ± 703.08	337.44 ± 141.09	<0.001 **
Cathepsin G (µg/mL)	6.79 ± 7.04	4.54 ± 6.28	0.080
LVV-H7 (ng/mL)	371.68 ± 190.52	194.48 ± 63.10	<0.001 **
Angiotensin II (pg/mL)	78.50 ± 105.78	87.81 ± 95.54	0.630

An independent t-test was used to compare continuous variables between the sepsis and non-sepsis groups. * $p < 0.05$; ** $p < 0.001$. ABG, arterial blood gas; ALT, alanine transaminase; aPTT, activated partial thromboplastin time; AST, aspartate transaminase; BUN, blood urea nitrogen; CRP, C-reactive protein; INR, international normalized ratio; LVV-H7, LVV-hemorphin 7; PCT, procalcitonin; P/F ratio, ratio of arterial oxygen partial pressure to fractional inspired oxygen; PT, prothrombin time; RBC, red blood cell count; SaO_2, arterial oxygen saturation; WBC, white blood cell.

As shown in Figure 2 and Table 2, the sepsis group had significantly higher levels of free Hb (15.27 vs. 9.75 mg/dL; $p < 0.001$), cathepsin D (849.19 vs. 337.44 ng/mL; $p < 0.001$), and LVV-H7 (371.68 vs. 194.48 ng/mL; $p < 0.001$) than the non-sepsis group. However, there were no significant differences observed in cathepsin G or angiotensin II levels between the two groups.

3.4. Potential Factors for Diagnosis of Sepsis, Shock, or Septic Shock in Critical Ill Patients with Acute Change of SOFA Score ≥ 2

Forward stepwise logistic regression identified variables with $p < 0.05$ as significant factors associated with sepsis classifications. Significant independent variables were CRP (odds ratio [OR] 1.086, 95% confidence interval [CI] 1.024–1.153; $p < 0.001$), aPTT (OR 0.999, 95% CI 0.999–1.163; $p < 0.05$), and WBC (OR 1.128, 95% CI 1.017–1.251; $p < 0.05$). After adjusting for other factors, LVV-H7 (adjusted OR [aOR] 1.009, 95% CI 1.005–1.013; $p < 0.001$) was a significant factor for differentiating sepsis classifications (Table 3).

The receiver operating characteristic (ROC) curves were calculated to predict sepsis using different biomarkers (Table 4 and Figure 4). The area under the curve (AUC) was 0.741 for LVV-H7, with the best cutoff point being 316.61 ng/mL (50.7% sensitivity and 100.0% specificity). The AUC was 0.738 for CRP, and a cutoff point of 5.85 mg/dL (72% sensitivity and 77.8% specificity. The AUC was 0.639 in WBC, with a cutoff point of 12.03 × 10^3/µL (53.3% sensitivity and 82.2% specificity). According to Youden's J statistic and AUC, LVV-H7 was the best biomarker for the diagnosing of sepsis, with high specificity and low sensitivity (Figure 4).

Table 3. Logistic regression models of sepsis classification factors.

Variables	β	SE	OR	95% CI		p-Value
				Lower	Upper	
LVV-H7	0.009	0.002	1.009	1.005	1.014	<0.001 **
CRP	0.083	0.030	1.086	1.024	1.153	0.006 *
aPTT	0.075	0.039	1.078	0.999	1.163	0.052
WBC	0.121	0.053	1.128	1.017	1.251	0.022 *

* $p < 0.05$; ** $p < 0.001$. Logistic regression model with forward stepwise selection for variables such as free hemoglobin, cathepsin D, cathepsin G, LVV-H7, angiotensin II, SOFA score, initial body temperature, MAP, Total bilirubin, CRP, aPTT, WBC, neutrophils, and lymphocytes. aPTT, activated partial thromboplastin time; CI, confidence interval; CRP, C-reactive protein; LVV-H7, LVV-hemorphin 7; OR, odds ratio; SE, standard error; WBC, white blood cell.

Table 4. ROC curves of different biomarkers of sepsis classifications (n = 120).

Biomarkers	Area	95% CI		p-Value	Cuff Point	Sensitivity	Specificity	Youden's J
		Lower	Upper					
LVV-H7	0.741	0.654	0.827	<0.001 **	316.61 (ng/mL)	0.507	1.000	0.507
CRP	0.738	0.647	0.830	<0.001 **	5.85 (mg/dL)	0.720	0.778	0.498
WBC	0.639	0.540	0.738	0.011 *	12.03 (10^3/μL)	0.533	0.822	0.356

Notes: Youden's J statistic: sensitivity + specificity −1. * $p < 0.05$; ** $p < 0.001$. CI, confidence interval; CRP, C-reactive protein; LVV-H7, LVV-hemorphin 7; ROC, receiver operating characteristic; WBC, white blood cell.

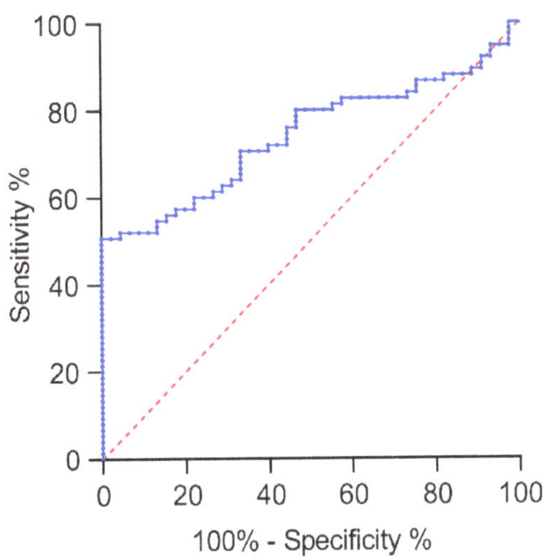

Figure 4. The receiver operating characteristic (ROC) curve for LVV-H7 in predicting sepsis. The blue solid line represents the ROC curve of LVV-H7 (LVV-hemorphin-7), illustrating its diagnostic performance. The red dashed line indicates the 50% area under the curve (AUC), serving as the reference line for random chance.

Forward stepwise logistic regression identified significant independent variables with $p < 0.05$ for predicting shock or septic shock. Among the 120 ICU patients, 81 experienced shock episodes within 48 h of ICU admission. Significant independent variables for predicting shock were Cathepsin G (aOR 0.917, 95% CI 0.848–0.991; $p < 0.05$), LVV-H7 (aOR 1.005, 95% CI 1.002–1.008; $p < 0.05$), SOFA score (OR 1.341, 95% CI 1.103–1.630; $p < 0.05$), and aPTT (OR 1.096, 95% CI 1.015–1.184; $p < 0.05$).

For predicting septic shock, among the 120 ICU patients, 40 were identified within 48 h of ICU admission. Significant independent variables for predicting septic shock were cathepsin D (aOR 1.001, 95% CI 1.000–1.002; $p < 0.05$), lactate (OR 1.182, 95% CI 1.054–1.324; $p < 0.05$), and CRP (OR 1.094, 95% CI 1.044–1.147; $p < 0.001$) (Table 4).

4. Discussion

This study is the first to evaluate Hb metabolism biomarkers, including free Hb, cathepsin D, cathepsin G, LVV-H7, and angiotensin II, in relation to sepsis classifications among critically ill patients with an acute increase in SOFA score ≥ 2.

Our findings demonstrate that critically ill definite and probable sepsis patients had higher levels of CFH, cathepsin G, cathepsin D, and LVV-H7 compared to possible sepsis patients and healthy controls, indicating potential metabolic interactions between immune cell lysosome enzymes and CFH during sepsis (Figure 5). These results align with previous research suggesting that CFH contributes to oxidative and endothelial injury in sepsis-related acute respiratory distress syndrome [21], a process well-documented in hemolysis and CFH release [4]. The CFH released into the blood increases the chances of cathepsin G originating from neutrophils and cathepsin D originating from macrophages digesting free Hb. Although cathepsin G levels were higher in the definite and probable sepsis group, it was not significantly different from those in the possible sepsis group, possibly because cathepsin G is released by neutrophils during inflammation and contributes to the immune response when the body experiences infection or inflammation [22,23]. Similarly, although cathepsin D levels were significantly higher in the definite and probable sepsis patients, they did not show a positive association in the logistic regression model for sepsis classification, possibly because cathepsin D is released from macrophages in severe acidic environments, leading to variable blood concentrations that may not accurately reflect sepsis status [24]. Despite the lack of association for cathepsins G and D in predicting sepsis classifications, these enzymes contribute to the production of downstream metabolites, such as LVV-H7, which was identified as a stable and predominant blood biomarker [12,25]. In this study, LVV-H7 was a significant factor for sepsis classifications (Table 3), although its levels may be influenced by comorbidities like cancer [26], obesity [27], and diabetes [28]. Here, malignancies, diabetes, and BMI calculations were investigated, and no differences in their distributions between the different sepsis groups were found. Notably, LVV-H7 demonstrated high specificity (100%) but low sensitivity (50.7%) for sepsis classifications, suggesting it may be more useful as an exclusionary marker rather than a primary diagnostic tool, similar to the role of D-dimer in ruling out pulmonary embolism [29].

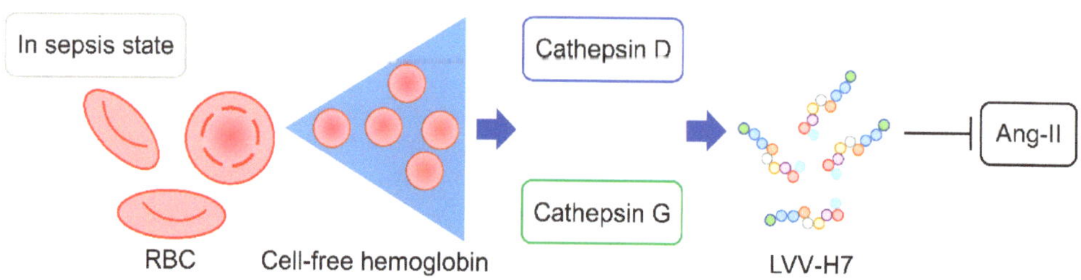

Figure 5. The signal transduction mechanism during sepsis. In sepsis state, RBC release cell-free hemoglobin due to infection, cathepsin D and cathepsin G degrade CFH, thereby releasing opioid peptides LVV-H7. The LVV-H7 binding to ANG-II, causing blood pressure decrease in patients with sepsis. CFH: cell-free hemoglobin; LVV-H7: LVV-hemorphin-7; RBC: red blood cells.

WBC was the traditional biomarker of systemic inflammation response syndrome as a previous diagnostic criterion of sepsis but was changed to the SOFA score at the 3rd International Sepsis Conference [18]. Here, WBC count was still a strong predictive factor for sepsis. In addition, the CRP level was also a traditional prognostic factor for sepsis in a clinical study [30] and was not included in any disease severity scoring system. Thus, the CRP level remained a substantial predictive factor for sepsis in this study.

In terms of clinical outcomes, the definite and probable sepsis group experienced significantly more shock episodes than the possible sepsis group, although no difference in the length of ICU stay, hospital stay, and survival. Our analysis of Hb metabolites (LVV-H7, cathepsin D, cathepsin G, and angiotensin II) in relation to shock and septic shock revealed that LVV-H7 could predict shock, while cathepsin G showed potential in reducing shock episodes (Table 5). The relationship between LVV-H7 and angiotensin II levels suggested dual mechanisms, with LVV-H7 potentially influencing blood pressure regulation through both angiotensin-converting enzyme inhibition and direct targeting of the angiotensin II type 1 receptor (AT1R) [14]. These mechanisms result in hypotension driven by LVV-H7, rather than by angiotensin II. Conversely, cathepsin G showed potential to predict less shock in multivariate logistic regression (Table 5). Cathepsin G traditionally acts as neutrophil angiotensin II-generating protease with functional control of blood pressure [31]. Cathepsin G may contribute to angiotensin II production in pig renal tissue [32], and these mechanisms may explain our clinical results that cathepsin G regulates the blood pressure with fewer shock episodes. Furthermore, cathepsin D showed potential to predict septic shock (Table 5). Septic shock has strict criteria of profound shock with lactic acidosis. Previous studies showed that cathepsin D is released more in acidic environments [24], suggesting an association with septic shock with lactic acidosis. Therefore, the lysosomal enzyme (cathepsin G and cathepsin D), CFH metabolites (LVV-H7), and angiotensin II may act as different indicators to guide clinical infection outcomes, including sepsis, shock, or septic shock.

Table 5. Logistic regression models of shock or septic shock factors.

Variables	β	SE	OR	95% CI		*p*-Value
				Lower	Upper	
For shock factors						
Cathepsin G	−0.087	0.040	0.917	0.848	0.991	0.030 *
LVV-H7	0.005	0.002	1.005	1.002	1.008	0.004 *
SOFA score	0.293	0.100	1.341	1.103	1.630	0.003 *
aPTT	0.092	0.039	1.096	1.015	1.184	0.019 *
For septic shock factors						
Cathepsin D	0.001	0.000	1.001	1.000	1.002	0.026 *
Lactate	0.167	0.058	1.182	1.054	1.324	0.004 *
CRP	0.090	0.024	1.094	1.044	1.147	<0.001 **

* $p < 0.05$; ** $p < 0.001$. Logistic regression model with forward stepwise selection for variables such as free hemoglobin, cathepsin D, cathepsin G, LVV-H7, Angiotensin II, SOFA score, gender, BMI, CCI, APACHE II score, lactate, oxygenation device status, albumin, CRP, PT, aPTT, INR, WBC, neutrophils, and lymphocytes. APACHE, Acute Physiology and Chronic Health Evaluation; aPTT, activated partial thromboplastin time; BMI, body mass index; CCI, Charlson comorbidity index; CI, confidence interval; CRP, C-reactive protein; INR, international normalized ratio; LVV-H7, LVV-hemorphin 7; OR, odds ratio; PT, prothrombin time, SE, standard error; SOFA, Sequential Organ Failure Assessment; WBC, white blood cell.

Procalcitonin is a highly sensitive diagnostic marker for sepsis and is also used to monitor the effectiveness of antibiotic therapy [33]. However, in our study, PCT was not associated with sepsis classifications. The possible reasons are as follows. Our study was designed to differentiate the causes of systemic inflammatory response syndrome (multiple organ and tissue hypoperfusion) and the hemoglobin metabolism status between definite, probable, and possible sepsis. Since all patients met the sepsis criteria, the possible sepsis group still exhibited high PCT levels. Another reason could be that we excluded cases where clinical cultures were collected 48 h after ICU admission. This exclusion might have affected the possible sepsis group, as some patients may have had insidious bacterial growth leading to high PCT levels. The exclusion was intended to prevent secondary infections from confounding the study results.

Some limitations of this study include the small sample size of the non-sepsis group. Although disease severity and comorbidities were similar between the sepsis and non-sepsis groups, the sample size in the non-sepsis and control groups, which was 50%

smaller than the sepsis group, may have reduced the statistical power and impacted the results of LVV-H7 as a less sensitive biomarker. Additionally, the SOFA score, a major diagnostic criterion for sepsis (acute change > 2), made it challenging for the non-sepsis and control groups to meet the inclusion criteria, resulting in exclusions based on these criteria. Increasing the control group size may enhance the statistical power for case–control study comparisons. Future studies should consider multicenter prospective designs enrolling sepsis and non-sepsis cases in a 1:1 ratio with SOFA > 2 as control groups. Secondly, in the sepsis group (n = 75), the diversity of organisms involved in sepsis may have affected the validation of sepsis biomarkers. Future comparisons between different classifications, such as Gram-positive vs. Gram-negative, multidrug-resistant organisms vs. non-resistant, and bacterial vs. viral infections, could help clarify the characteristics of new sepsis biomarkers. Additionally, we did not test procalcitonin more than once in the non-sepsis group for a series of comparisons. Procalcitonin is more sensitive to bacterial infection and suitable for new biomarker comparisons. We suggest routine procalcitonin testing as a follow-up measure in future sepsis studies.

5. Conclusions

LVV-H7, cathepsin D, and cathepsin G levels serves as potential biomarkers for sepsis classifications and shock episodes in critically ill patients with acute changes in the SOFA score ≥ 2. LVV-H7 may help exclude definite and probable sepsis when levels are below normal. Cathepsin G may assist in blood pressure control and reduce shock incidence, while cathepsin D is more active and associated with septic shock. Further multicenter prospective studies that enroll other non-sepsis control groups are needed to validate the clinical utility of cathepsin D, cathepsin G, and LVV-H7 in differentiating sepsis and septic shock.

Author Contributions: W.-L.S. and Y.-K.W. conceived and designed the study and obtained research funding. W.-L.S., Y.-T.C., Y.-K.W. and H.-W.C. supervised the study, collected the data, and critically revised the manuscript. W.-L.S., H.-W.C., H.-C.C., I.-H.C. and Y.-T.C. provided statistical advice on the study design and analyzed the data, while W.-L.S., H.-W.C., Y.-K.W., Y.-T.C., H.-C.C. and I.-H.C. conducted data analysis, data interpretation, and manuscript preparation. W.-L.S. and H.-W.C. aired the Data Oversight Committee. W.-L.S. drafted the manuscript and all authors contributed substantially to its revision. Y.-T.C. and Y.-K.W. took responsibility for the entire manuscript. All authors have read and agreed to the published version of the manuscript.

Funding: This work was supported by grants from the Buddhist Tzu Chi Medical Foundation [TCMF-CP 112-02] and Taipei Tzu Chi Hospital [TCRD-TPE-112-03]. The funding body had no role in the conceptualization, design, data collection, analysis, decision to publish, or preparation of the manuscript.

Institutional Review Board Statement: The study was conducted in accordance with the Declaration of Helsinki, and approved by the Institutional Review Board of Taipei Tzu Chi Hospital, Buddhist Tzu Chi Medical Foundation, New Taipei City, Taiwan on March 16, 2020 (Protocol No.: 08-P-134).

Informed Consent Statement: Informed consent was obtained from all subjects involved in the study.

Data Availability Statement: The data supporting the findings of this study are available from the Taipei Tzu Chi Hospital, Buddhist Tzu Chi Medical Foundation. Restrictions may apply to the availability of data used under the license of the current study because they are not publicly available. However, the data are available from the authors upon reasonable request and with permission from Taipei Tzu Chi Hospital and the Buddhist Tzu Chi Medical Foundation.

Acknowledgments: The authors appreciate the assistance of the Biobank of Taipei Tzu Chi Hospital in the collection and processing of clinical specimens. We acknowledge Chih-Yu Chan, the Division of Pulmonary and Critical Care Medicine, Department of Internal Medicine, Taipei Tzu Chi Hospital, Buddhist Tzu Chi Medical Foundation, New Taipei, Taiwan, for conducting ELISA to detect cell-free hemoglobin, cathepsin D, cathepsin G, LVV-H7, and angiotensin II.

Conflicts of Interest: The authors declare that they have no competing interests.

List of Abbreviations

aHR, adjusted hazard ratio; aOR, adjusted odds ratio; aPTT, activated partial thromboplastin clotting time; AUC: area under the curve; BMI, body mass index; BT: body temperature; BUN, blood urea nitrogen; CCI, Charlson comorbidity index; CI, confidence interval; CFH: cell-free hemoglobin; CRP, C-reactive protein; ELISA, enzyme-linked immunosorbent assay; GCS, Glasgow Coma Scale; Hb, hemoglobin; LVV-H7, LVV-hemorphin-7; MAP, Mean arterial pressure; PCT, procalcitonin; qSOFA, quick sepsis-related organ failure assessment; RBC: red blood cells; SOFA, Sequential Organ Failure Assessment Score; SpO_2, oxygen saturation of pulse oximeter; WBC: white blood cells.

References

1. Jansma, G.; de Lange, F.; Kingma, W.P.; Vellinga, N.A.; Koopmans, M.; Kuiper, M.A.; Boerma, E.C. 'Sepsis-related anemia' is absent at hospital presentation; a retrospective cohort analysis. *BMC Anesthesiol.* **2015**, *15*, 55. [CrossRef] [PubMed]
2. Nissenson, A.R.; Dylan, M.L.; Griffiths, R.I.; Yu, H.T.; Dubois, R.W. Septicemia in patients with ESRD is associated with decreased hematocrit and increased use of erythropoietin. *Clin. J. Am. Soc. Nephrol.* **2006**, *1*, 505–510. [CrossRef] [PubMed]
3. van Beest, P.A.; Hofstra, J.J.; Schultz, M.J.; Boerma, E.C.; Spronk, P.E.; Kuiper, M.A. The incidence of low venous oxygen saturation on admission to the intensive care unit: A multi-center observational study in The Netherlands. *Crit. Care* **2008**, *12*, R33. [CrossRef]
4. Effenberger-Neidnicht, K.; Hartmann, M. Mechanisms of Hemolysis During Sepsis. *Inflammation* **2018**, *41*, 1569–1581. [CrossRef]
5. Larsen, R.; Gozzelino, R.; Jeney, V.; Tokaji, L.; Bozza, F.A.; Japiassu, A.M.; Bonaparte, D.; Cavalcante, M.M.; Chora, A.; Ferreira, A.; et al. A central role for free heme in the pathogenesis of severe sepsis. *Sci. Transl. Med.* **2010**, *2*, 51ra71. [CrossRef] [PubMed]
6. Dutra, F.F.; Bozza, M.T. Heme on innate immunity and inflammation. *Front. Pharmacol.* **2014**, *5*, 115. [CrossRef]
7. Adamzik, M.; Hamburger, T.; Petrat, F.; Peters, J.; de Groot, H.; Hartmann, M. Free hemoglobin concentration in severe sepsis: Methods of measurement and prediction of outcome. *Crit. Care* **2012**, *16*, R125. [CrossRef]
8. Janz, D.R.; Bastarache, J.A.; Peterson, J.F.; Sills, G.; Wickersham, N.; May, A.K.; Roberts, L.J., 2nd; Ware, L.B. Association between cell-free hemoglobin, acetaminophen, and mortality in patients with sepsis: An observational study. *Crit. Care Med.* **2013**, *41*, 784–790. [CrossRef] [PubMed]
9. Starkey, P.M.; Barrett, A.J. Human cathepsin G. Catalytic and immunological properties. *Biochem. J.* **1976**, *155*, 273–278. [CrossRef]
10. Korkmaz, B.; Horwitz, M.S.; Jenne, D.E.; Gauthier, F. Neutrophil elastase, proteinase 3, and cathepsin G as therapeutic targets in human diseases. *Pharmacol. Rev.* **2010**, *62*, 726–759. [CrossRef]
11. Diment, S.; Leech, M.S.; Stahl, P.D. Cathepsin D is membrane-associated in macrophage endosomes. *J. Biol. Chem.* **1988**, *263*, 6901–6907. [CrossRef] [PubMed]
12. Mielczarek, P.; Hartman, K.; Drabik, A.; Hung, H.Y.; Huang, E.Y.; Gibula-Tarlowska, E.; Kotlinska, J.H.; Silberring, J. Hemorphins-From Discovery to Functions and Pharmacology. *Molecules* **2021**, *26*, 3879. [CrossRef] [PubMed]
13. Dejouvencel, T.; Feron, D.; Rossignol, P.; Sapoval, M.; Kauffmann, C.; Piot, J.M.; Michel, J.B.; Fruitier-Arnaudin, I.; Meilhac, O. Hemorphin 7 reflects hemoglobin proteolysis in abdominal aortic aneurysm. *Arterioscler. Thromb. Vasc. Biol.* **2010**, *30*, 269–275. [CrossRef]
14. Ali, A.; Palakkott, A.; Ashraf, A.; Al Zamel, I.; Baby, B.; Vijayan, R.; Ayoub, M.A. Positive Modulation of Angiotensin II Type 1 Receptor-Mediated Signaling by LVV-Hemorphin-7. *Front. Pharmacol.* **2019**, *10*, 1258. [CrossRef] [PubMed]
15. Calandra, T.; Cohen, J.; FRCP for the International Sepsis Forum Definition of Infection in the ICU Consensus Conference. The international sepsis forum consensus conference on definitions of infection in the intensive care unit. *Crit. Care Med.* **2005**, *33*, 1538–1548. [CrossRef]
16. Rhee, C.; Jones, T.M.; Hamad, Y.; Pande, A.; Varon, J.; O'Brien, C.; Anderson, D.J.; Warren, D.K.; Dantes, R.B.; Epstein, L.; et al. Prevalence, Underlying Causes, and Preventability of Sepsis-Associated Mortality in US Acute Care Hospitals. *JAMA Netw. Open* **2019**, *2*, e187571. [CrossRef]
17. Mellhammar, L.; Elen, S.; Ehrhard, S.; Bouma, H.; Ninck, L.; Muntjewerff, E.; Wunsch, D.; Bloos, F.; Malmstrom, E.; Linder, A. New, Useful Criteria for Assessing the Evidence of Infection in Sepsis Research. *Crit. Care Explor.* **2022**, *4*, e0697. [CrossRef]
18. Singer, M.; Deutschman, C.S.; Seymour, C.W.; Shankar-Hari, M.; Annane, D.; Bauer, M.; Bellomo, R.; Bernard, G.R.; Chiche, J.D.; Coopersmith, C.M.; et al. The Third International Consensus Definitions for Sepsis and Septic Shock (Sepsis-3). *JAMA* **2016**, *315*, 801–810. [CrossRef]
19. Evans, L.; Rhodes, A.; Alhazzani, W.; Antonelli, M.; Coopersmith, C.M.; French, C.; Machado, F.R.; McIntyre, L.; Ostermann, M.; Prescott, H.C.; et al. Surviving Sepsis Campaign: International Guidelines for Management of Sepsis and Septic Shock 2021. *Crit. Care Med.* **2021**, *49*, e1063–e1143. [CrossRef]
20. Charlson, M.E.; Pompei, P.; Ales, K.L.; MacKenzie, C.R. A new method of classifying prognostic comorbidity in longitudinal studies: Development and validation. *J. Chronic Dis.* **1987**, *40*, 373–383. [CrossRef]
21. Janz, D.R.; Ware, L.B. The role of red blood cells and cell-free hemoglobin in the pathogenesis of ARDS. *J. Intensive Care* **2015**, *3*, 20. [CrossRef] [PubMed]
22. Mantovani, A.; Cassatella, M.A.; Costantini, C.; Jaillon, S. Neutrophils in the activation and regulation of innate and adaptive immunity. *Nat. Rev. Immunol.* **2011**, *11*, 519–531. [CrossRef]

23. Zamolodchikova, T.S.; Tolpygo, S.M.; Svirshchevskaya, E.V. Cathepsin G-Not Only Inflammation: The Immune Protease Can Regulate Normal Physiological Processes. *Front. Immunol.* **2020**, *11*, 411. [CrossRef] [PubMed]
24. Yadati, T.; Houben, T.; Bitorina, A.; Shiri-Sverdlov, R. The Ins and Outs of Cathepsins: Physiological Function and Role in Disease Management. *Cells* **2020**, *9*, 1679. [CrossRef]
25. Fruitier, I.; Garreau, I.; Piot, J.M. Cathepsin D is a good candidate for the specific release of a stable hemorphin from hemoglobin in vivo: VV-hemorphin-7. *Biochem. Biophys. Res. Commun.* **1998**, *246*, 719–724. [CrossRef] [PubMed]
26. Cohen, M.; Fruitier-Arnaudin, I.; Sauvan, R.; Birnbaum, D.; Piot, J.M. Serum levels of Hemorphin-7 peptides in patients with breast cancer. *Clin. Chim. Acta* **2003**, *337*, 59–67. [CrossRef]
27. Maraninchi, M.; Feron, D.; Fruitier-Arnaudin, I.; Begu-Le Corroller, A.; Nogueira, J.P.; Mancini, J.; Valero, R.; Piot, J.M.; Vialettes, B. Serum hemorphin-7 levels are decreased in obesity. *Obesity* **2013**, *21*, 378–381. [CrossRef]
28. Fruiter, A., II; Cohen, M.M.; Nervi, S.S.; Bordenave, S.S.; Sannier, F.F.; Piot, J.M. Reduced level of opioid peptides, hemorphin-7 peptides, in serum of diabetic patients. *Diabetes Care* **2003**, *26*, 2480. [CrossRef]
29. Kearon, C.; de Wit, K.; Parpia, S.; Schulman, S.; Afilalo, M.; Hirsch, A.; Spencer, F.A.; Sharma, S.; D'Aragon, F.; Deshaies, J.F.; et al. Diagnosis of Pulmonary Embolism with d-Dimer Adjusted to Clinical Probability. *N. Engl. J. Med.* **2019**, *381*, 2125–2134. [CrossRef]
30. Koozi, H.; Lengquist, M.; Frigyesi, A. C-reactive protein as a prognostic factor in intensive care admissions for sepsis: A Swedish multicenter study. *J. Crit. Care* **2020**, *56*, 73–79. [CrossRef]
31. Tonnesen, M.G.; Klempner, M.S.; Austen, K.F.; Wintroub, B.U. Identification of a human neutrophil angiotension II-generating protease as cathepsin G. *J. Clin. Investig.* **1982**, *69*, 25–30. [CrossRef] [PubMed]
32. Rykl, J.; Thiemann, J.; Kurzawski, S.; Pohl, T.; Gobom, J.; Zidek, W.; Schlüter, H. Renal cathepsin G and angiotensin II generation. *J. Hypertens.* **2006**, *24*, 1797–1807. [CrossRef] [PubMed]
33. Vijayan, A.L.; Vanimaya; Ravindran, S.; Saikant, R.; Lakshmi, S.; Kartik, R.; G, M. Procalcitonin: A promising diagnostic marker for sepsis and antibiotic therapy. *J. Intensive Care* **2017**, *5*, 51. [CrossRef] [PubMed]